Collins

Cambridge IGCSE™
Combined Science

WORKBOOK

Gurinder Chadha, Heidi Foxford, Aidan Gill,
Mike Smith and Chris Sunley

William Collins' dream of knowledge for all began with the publication of his first book in 1819.
A self-educated mill worker, he not only enriched millions of lives, but also founded a flourishing publishing house. Today, staying true to this spirit, Collins books are packed with inspiration, innovation and practical expertise.
They place you at the centre of a world of possibility and give you exactly what you need to explore it.

Collins. Freedom to teach.

Published by Collins
An imprint of HarperCollins*Publishers*
The News Building, 1 London Bridge Street, London, SE1 9GF, UK

HarperCollins*Publishers*
Macken House, 39/40 Mayor Street Upper, Dublin 1, D01 C9W8, Ireland

Browse the complete Collins catalogue at
collins.co.uk

© HarperCollins*Publishers* Limited 2025

10 9 8 7 6 5 4 3 2 1

ISBN 978-0-00-875857-8

All rights reserved. No part of this publication may be reproduced, stored in a retrieval system, or transmitted in any form by any means, electronic, mechanical, photocopying, recording or otherwise, without the prior written permission of the Publisher or a licence permitting restricted copying in the United Kingdom issued by the Copyright Licensing Agency Ltd, 5th Floor, Shackleton House, 4 Battle Bridge Lane, London SE1 2HX.

Without limiting the author's and publisher's exclusive rights, any unauthorised use of this publication to train generative artificial intelligence (AI) technologies is expressly prohibited. HarperCollins also exercise their rights under Article 4(3) of the Digital Single Market Directive 2019/790 and expressly reserve this publication from the text and data mining exception.

British Library Cataloguing-in-Publication Data
A catalogue record for this publication is available from the British Library.

Authors: **Gurinder Chadha**, **Heidi Foxford**, **Aidan Gill**, **Mike Smith** and **Chris Sunley**
Expert reviewers: **Gauri Tendulkar**, **Frank Akrofi** and **Samuel Yeboah**
Publisher: **Catherine Martin**
Product managers: **Jessica Ashdale** and **Saaleh Patel**

Copyeditor: **Gemma Young**
Proofreader: **Arlo Porter**
Answer checker: **Dr Sarah Ryan**
Cover designer: **Gordon MacGilp**
Cover artwork: **Ann Paganuzzi**
Internal designer and illustrator: **PDQ Media**
Typesetter: **PDQ Media**
Production controller: **Alhady Ali**

Printed and bound in the UK by Martins the Printers

This book is produced from independently certified FSC™ paper to ensure responsible forest management. For more information visit:
www.harpercollins.co.uk/green

Acknowledgements
With thanks to the following teachers for reviewing materials and providing valuable feedback: **Frank Akrofi**, The Roman Ridge School; **Dr Raul Balbuena**, Tama Rama Intercultural School; **Dr Rahul Sharma**, IRA Global School; **Dániel Szücs**, International School of Budapest; **Gauri Tendulkar**, JBCN International School; **Samuel Yeboah**, AVES International Academy; and with thanks to the following teachers who provided feedback during the development stages: **Shalini Reddy**, Manthan International School; **Sejal Vasarkar**, SVKM JV Parekh International School.

Cambridge International Education material in this publication is reproduced under licence and remains the intellectual property of Cambridge University Press & Assessment.

This text has not been through the endorsement process for the Cambridge Pathway. Any references or materials related to answers, grades, papers or examinations are based on the opinion of the author(s). The Cambridge International Education syllabus or curriculum framework associated assessment guidance material and specimen papers should always be referred to for definitive guidance.

The publishers gratefully acknowledge the permission granted to reproduce the copyright material in this book. Every effort has been made to trace copyright holders and to obtain their permission for the use of copyright material. The publishers will gladly receive any information enabling them to rectify any error or omission at the first opportunity.

Photographs
p 42 Shutterstock/Schira; p 185 Shutterstock/Triff; p 208 Shutterstock/Grigvovan

Biology

B1 Characteristics of living organisms — 7
Characteristics of living organisms — 7

B2 Cells — 8
Cell structure — 8
Size of specimens — 9

B3 Movement into and out of cells — 11
Diffusion — 11
Osmosis — 13
SUPPLEMENT Active transport — 15

B4 Biological molecules — 16
Biological molecules — 16

B5 Enzymes — 18
Enzymes — 18

B6 Plant nutrition — 22
Photosynthesis — 22
Leaf structure — 25

B7 Human nutrition — 26
Diet — 26
Digestive system — 28
Digestion — 30

B8 Transport in plants — 32
Xylem and phloem — 32
Water uptake — 33
Transpiration — 34

B9 Transport in animals — 36
Circulatory systems — 36
Heart — 36
Blood vessels — 39
Blood — 41

B10 Diseases and immunity — 44
Diseases and immunity — 44

B11 Gas exchange in humans — **48**
Gas exchange in humans — 48

B12 Respiration — **50**
Respiration — 50

B13 Drugs — **51**
Drugs — 51

B14 Reproduction — **52**
Sexual reproduction in plants — 52
Sexual reproduction in humans — 54

B15 Organisms and their environment — **56**
Energy flow — 56
Food chains and food webs — 56
Carbon cycle — 59

B16 Human influences on ecosystems — **60**
Habitat destruction — 60
Conservation — 61

B1 Characteristics of living organisms

Characteristics of living organisms

Student's Book pages 12–14 | Syllabus learning objective B1.1.1

1 a The boxes on the left below show some of the characteristics of living organisms.

The boxes on the right show some descriptions of characteristics.

Draw lines to link each characteristic to its description. Draw **four** lines.

characteristic	description
movement	ability to detect and respond to changes in the environment
reproduction	breakdown of nutrient molecules to release energy
respiration	change of position or place
sensitivity	increase in size or dry mass
	making more of the same kind of organism

[4]

b Circle the characteristics that are shown by **all** living organisms.

digestion   excretion   growth

nutrition   photosynthesis   sight   transpiration

[3]

B2 Cells
Cell structure

Student's Book pages 18–26 | Syllabus learning objectives B2.1.1–B2.1.5; SUPPLEMENT B2.1.6

1 The diagram shows a plant cell.

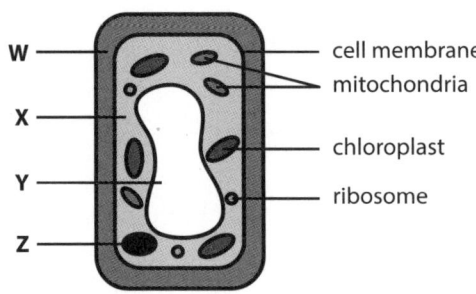

a State the names of structures **W**, **X**, **Y** and **Z**.

W .... Y ....

X .... Z .... [4]

b Identify **three** structures shown in the diagram that are **not** found in animal cells.

1 .... 3 .... [3]

2 ....

c Describe the functions of the following cell structures.

cell membrane ....

mitochondria ....

ribosomes .... [3]

d State the name of **one** structure found in bacterial cells that is **not** shown in the diagram.

.... [1]

2 Complete the sentence about cells using terms from the list. Each term can be used once, more than once or not at all.

       organ systems   **organisms**   **organs**   **tissues**

Similar cells are grouped together to form .... which, in turn, can be grouped together to form .... [2]

3 State the functions of the following specialised cells:

SUPPLEMENT

root hair cell ....

palisade mesophyll cell ....

red blood cell .... [3]

Size of specimens

Student's Book pages 26–27 | Syllabus learning objectives B2.2.1–B2.2.2; **SUPPLEMENT** B2.2.3

1 a Complete the formula for magnification.

magnification = ─────────────── [2]

b A student uses a microscope to take an image of a cell.

The length of the cell image is 60 mm.

The magnification of the image is ×200.

Calculate the actual length of the cell in mm.

.... mm [3]

c State the actual length of the cell in micrometres (μm).

SUPPLEMENT

.... μm [1]

TIP
When doing a calculation, always show your working. That way, even if your final answer is incorrect, you may still gain some marks.

B3 Movement into and out of cells
Diffusion

Student's Book pages 34–38 | Syllabus learning objectives B3.1.1–B3.1.3; **SUPPLEMENT** B3.1.4

1 Which statement describes diffusion? Circle the correct letter.

   A net movement of particles from a region of high concentration to another region of high concentration

   B net movement of particles from a region of high concentration to a region of low concentration

   C net movement of particles from a region of low concentration to another region of low concentration

   D net movement of particles from a region of low concentration to a region of high concentration [1]

2 Define the term net movement.

....

....
[1]

3 Describe the importance of the diffusion of oxygen across cell membranes in animal cells.

....

....

....

....
[3]

TIP
Look at the number of marks available and try to make at least that number of distinct points or steps in your answer. In this case, you are looking for at least three points.

4 SUPPLEMENT

Some students investigate diffusion using the apparatus shown in the diagram below.

They cut a small cube of beetroot of size 1 cm³, which they wash and then place in clean water.

As the coloured substance diffuses out of the beetroot, it turns the water pink.

The students time how long it takes to turn the water a certain shade of pink. This indicates the rate of diffusion.

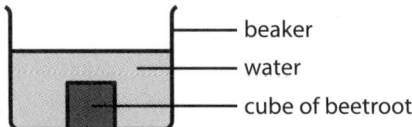

The students want to investigate how surface area affects the rate of diffusion.

They do this by repeating the experiment several times. Each time they start with a fresh 1 cm³ cube of beetroot, but cut each cube into a different number of smaller pieces.

a Identify **two** variables the students should control.

1 ....

2 .... [2]

b Suggest **one** way the students can make sure the water is the same shade of pink each time when they stop timing.

.... [1]

c Suggest the conclusion that the students will make from their results.

.... [1]

TIP
A question with the command word 'suggest' means you need to have a go at **applying** your knowledge and understanding to answer the question.

Osmosis

Student's Book pages 38–43 | Syllabus learning objectives B3.2.1–B3.2.3; SUPPLEMENT B3.2.4–B3.2.5

1 Complete the sentence about water using terms from the list. Each term can be used once, more than once or not at all.

    respiration   **impermeable**   **osmosis**   **permeable**

Water diffuses in and out of cells by .... through their partially .... membranes. [2]

2 Some students investigated osmosis. They cut identical blocks from a potato and measured their lengths. They left the blocks in three different solutions as shown in the diagram. After 30 minutes, they removed the blocks and measured their lengths again.

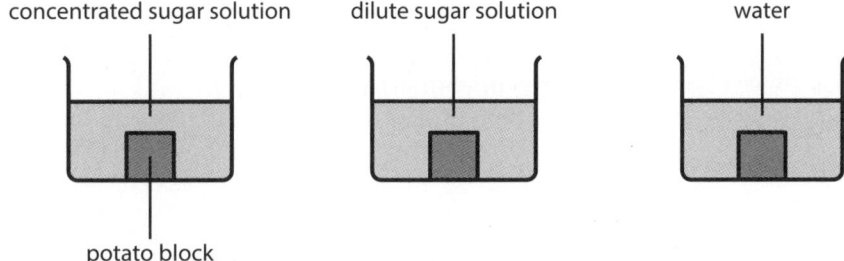

The students found that one block increased in length, one decreased and one stayed the same.

a Place ticks (✓) in the table to show their results.

Place **one** tick in **each** row.

	Increased in length	Decreased in length	Stayed the same length
Concentrated sugar solution			
Dilute sugar solution			
Water			

[2]

b Apart from the length of the potato blocks, state **one other** feature that the students could have measured to show the effects of different concentrations.

.... [1]

3 Define osmosis. Refer to water potential in your answer.

SUPPLEMENT

....

....

.... [3]

4 The diagram shows a turgid plant cell. In the space alongside, draw the same cell when it is **plasmolysed**.

SUPPLEMENT

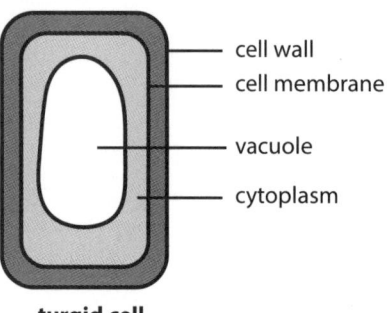

turgid cell   plasmolysed cell

[2]

Active transport

Student's Book pages 43–44 | Syllabus learning objectives
SUPPLEMENT B3.3.1–B3.3.2

1 Describe **two** ways in which the process of active transport is different from diffusion.

1 ....

....

2 ....

....

[2]

> **TIP**
> Make sure that your answer is as clear as possible. In this case, try to write 'active transport' or 'diffusion' instead of 'it' so that there is no confusion about what you are referring to.

2 The root hair cells of plants use active transport to take in mineral ions from the soil.

Explain why root hair cells need to use active transport to take in mineral ions from the soil.

....

....

....

[2]

> **TIP**
> The command word 'explain' means that you should give a reason or reasons in your answer.

B4 Biological molecules
Biological molecules

Student's Book pages 50–53 | Syllabus learning objectives B4.1.1–B4.1.3

1 (Circle) the elements that all carbohydrates, fats and proteins contain.

    calcium   carbon   hydrogen   iron

    magnesium   nitrogen   oxygen   sulfur

[3]

2 The boxes on the left show some large biological molecules. The boxes on the right show some smaller molecules.

Draw lines to link each large molecule to the smaller molecules it is made from. Draw **four** lines.

large biological molecules	smaller molecules
fats and oils	amino acids
proteins	fatty acids
	glucose
starch, glycogen, cellulose	glycerol

[4]

3 The table below shows the details of some tests for food molecules. Complete the table.

Food being tested for	Description of test	Positive result
Starch		colour changes from brown to blue-black
Reducing sugars, e.g. glucose	heat with Benedict's solution	
Proteins	add biuret solution	
	mix with ethanol and then add the solution to water	forms a cloudy white emulsion

[4]

B5 Enzymes
Enzymes

Student's Book pages 58–64 | Syllabus learning objectives B5.1.1–B5.1.2; **SUPPLEMENT** B5.1.3–B5.1.6

1 Place ticks (✓) in the boxes to show **three** correct statements about enzymes.

Enzymes act as biological catalysts.	
Enzymes act only in the digestive system.	
Enzymes are carbohydrates.	
Enzymes are fats.	
Enzymes are proteins.	
Enzymes maintain metabolic reaction rates.	
Enzymes reduce metabolic reaction rates.	

[3]

2 A student investigates the effect of changing pH on the activity of an enzyme.

The graph shows their results.

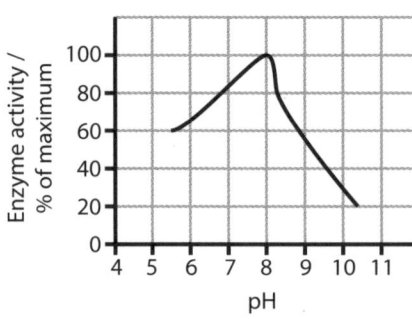

Describe the effect of changing pH on the enzyme shown in the graph.

TIP
The command word here, 'describe', means that you just have to put into words what the graph shows. You **do not** have to **explain** any reasons why the graph is this shape.

....

....

....

.... [3]

3 **SUPPLEMENT** The diagram shows the stages in an enzyme-catalysed reaction.

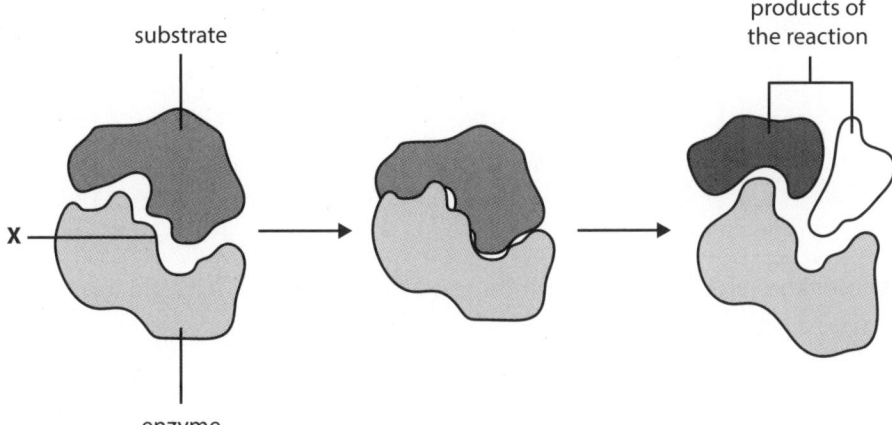

a Identify the area of the enzyme labelled **X**.

.... [1]

b Describe and explain the process of enzyme action as shown in the diagram.

....

....

....

.... [4]

> **TIP**
> Use the number of marks and the number of answer lines as a guide to the length of your answer. Do not worry if you do not always use all the answer lines, but you should try to make at least as many different points as there are marks.

c The enzyme shown in the diagram is only able to catalyse the breakdown of the substrate shown.

Explain why it **cannot** also catalyse the breakdown of other substances.

....

.... [2]

4 **SUPPLEMENT** The graph shows the effect of changes in temperature on the rate of reaction of an enzyme-catalysed reaction.

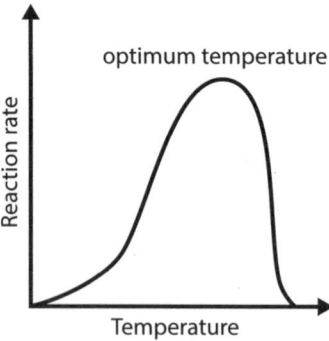

a Explain the shape of the graph as the temperature increases **up to** the optimum temperature.

....

....

.... [3]

b As the temperature increases above the optimum temperature, the enzyme starts to denature.

Explain why this affects the shape of the graph **above** the optimum temperature.

....

....

.... [3]

5 Which set of conditions would **not** start to denature an enzyme? Circle the correct letter.

SUPPLEMENT

A pH above optimum pH, temperature at optimum temperature

B pH at optimum pH, temperature below optimum temperature

C pH at optimum pH, temperature above optimum temperature

D pH below optimum pH, temperature at optimum temperature [1]

B6 Plant nutrition
Photosynthesis

Student's Book pages 70–75 | Syllabus learning objectives B6.1.1–B6.1.4; SUPPLEMENT B6.1.5–B6.1.8

1 a Complete the sentences about plant nutrition using words from the list. Each word can be used once, more than once or not at all.

> glucose   light   minerals
>
> photosynthesis   respiration   soil

Plants are able to make their own food using energy from ....
This process is called .... [2]

b When do plants make their own food? Circle the correct letter.

  A day and night

  B only at night

  C only in the day

  D only in the day if there are no clouds in the sky [1]

c Complete the word equation for photosynthesis.

.... + water $\xrightarrow[\text{chlorophyll}]{\text{sunlight}}$ .... + oxygen   [2]

d State the balanced symbol equation for photosynthesis.

SUPPLEMENT

.... [2]

e Photosynthesis requires a green pigment called chlorophyll.

State the name of the structures inside plant cells where chlorophyll is found.

.... [1]

f Which statement best describes the function of chlorophyll? Circle the correct letter.

SUPPLEMENT

A   It absorbs light energy and transfers it into chemical energy.

B   It absorbs light energy to keep the plant warm.

C   It makes the plant green and attractive to insects.

D   It transfers chemical energy into carbon dioxide. [1]

2 a A student keeps a variegated plant (a plant with green and white leaves) in a dark cupboard for two days. On the third day the plant is moved into bright sunlight and watered. At the end of the third day the leaf is tested for the presence of starch.

Predict the results of the starch test by completing the table.

Part of leaf	Colour change with starch test
Green	
White	

[2]

b Explain why the plant was kept in the dark cupboard for two days.

....

.... [2]

3 What is the colour of hydrogencarbonate indicator in water containing a high concentration of dissolved carbon dioxide? Circle the correct letter.

A   orange

B   purple

C   red

D   yellow [1]

4 The graph shows how carbon dioxide concentration affects the rate of photosynthesis.

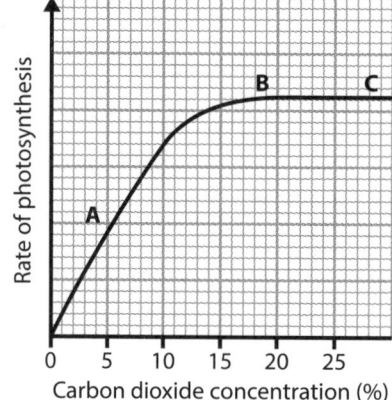

Explain the shape of the curve shown in the graph between points **A**, **B** and **C**.

....

....

....

.... [2]

Leaf structure

Student's Book page 76 | Syllabus learning objective B6.2.1

1 The diagram shows the internal structure of a leaf. Use terms from the list to complete the labels on the diagram. Each term can be used once, more than once or not at all.

guard cell   **lower epidermis**   **palisade mesophyll**   **phloem**

spongy mesophyll   **stoma**   **upper epidermis**   **xylem**

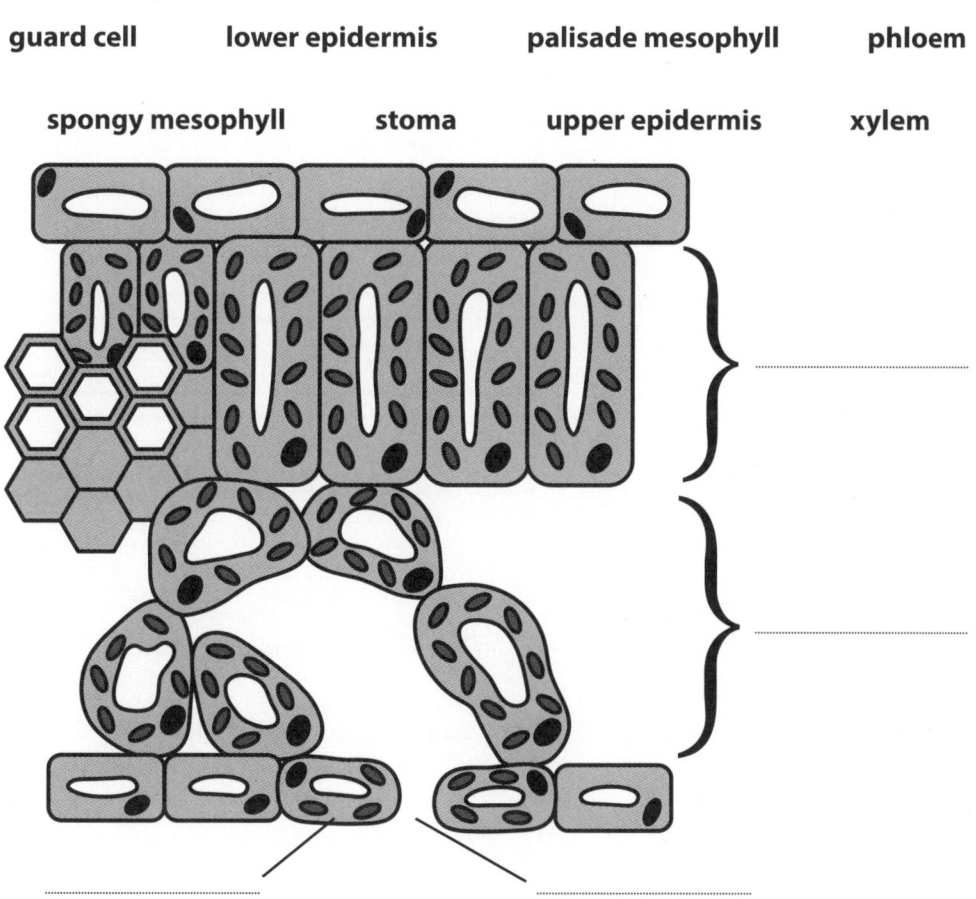

[4]

B7 Human nutrition
Diet

Student's Book pages 82–84 | Syllabus learning objectives B7.1.1–B7.1.2

1 Draw **one** line to match each food group with the reason it is important.

food group	reason it is important
carbohydrate	to store energy and maintain body temperature
fibre	for the movement of food by peristalsis
fats and oils	to keep skin, bones and teeth healthy
	for respiration

[3]

2 Which statement best describes a balanced diet? Circle the correct letter.

A a diet that contains all the nutrients in the recommended proportions

B a diet that contains no fat or sugar

C a diet that contains similar percentages of each food group

D a diet that has a very high percentage of fruit and vegetables

[1]

3 The table shows guidance for recommended percentages of different food groups in a balanced diet, and the percentages of each food group eaten by three students.

Food group	Recommended percentage	Student A	Student B	Student C
Fruit and vegetables	33	37	10	32
Starchy foods	33	34	39	34
Dairy products	15	11	14	13
Foods high in protein	12	15	12	18
Foods high in fat or sugar	7	3	25	3

a State which student has the **least** balanced diet. .... [1]

b State **two** reasons for your answer.

1 ....

2 .... [2]

c One student is an athlete and has a diet high in white meat, fish, lentils and eggs.

  i State which student is most likely to be the athlete. .... [1]

  ii Explain why the athlete regularly eats foods such as white meat, fish, lentils and eggs.

....

.... [2]

4 Anaemia is a deficiency disease in which the number of red blood cells in the body is reduced.

a State the name of the vitamin or mineral that is deficient in anaemia.

.... [1]

b Explain why people with anaemia often feel tired and short of breath.

....

.... [2]

Digestive system

Student's Book pages 85–86 | Syllabus learning objectives B7.2.1–B7.2.2

1 The diagram shows the human digestive system.

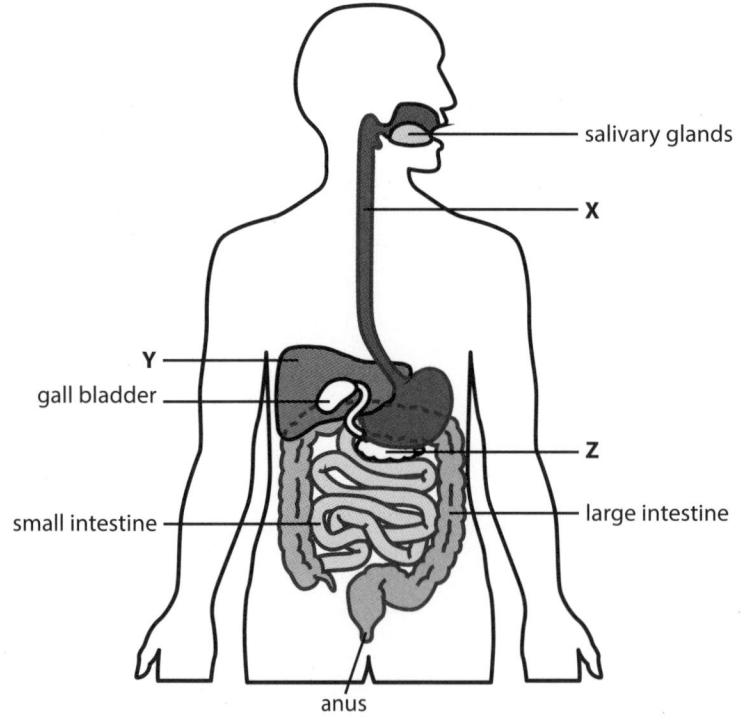

a State the names of structures **X**, **Y** and **Z**.

X ....

Y ....

Z .... [3]

b The small intestine has two parts, each with a slightly different function. State the **two** parts of the small intestine.

1 .... 2 .... [2]

c Describe the function in the digestive system of the structure labelled **Z**.

.... [1]

2 The table gives information on processes that take place in different parts of the digestive system.

Complete the table to provide the missing information.

Name of process	Where it takes place	Description
Ingestion		taking food or drink into the body
Absorption	intestines	
	alimentary canal	removal of undigested food from the body (faeces)

[3]

3 An inflamed gall bladder can cause the bile duct to become blocked. This condition can cause nausea, vomiting and abdominal pain due to the build-up of undigested fats.

Explain why fats might **not** get broken down in a person with an inflamed gall bladder.

....

.... [1]

Digestion

Student's Book pages 87–89 | Syllabus learning objectives B7.3.1–B7.3.2; **SUPPLEMENT** B7.3.3–B7.3.7

1 Complete the sentence about physical digestion using words from the list. Each word can be used once, more than once or not at all.

> chemical   larger   molecules
>
> physical   size   smaller

Physical digestion is the breakdown of food into .... pieces without any .... change to the food ....

[3]

2 Explain why physical digestion is important.

SUPPLEMENT

....

....

[2]

3 Complete the sentence about chemical digestion using terms from the list. Each term can be used once, more than once or not at all.

> alimentary canal   chewing   enzymes
>
> large intestine   gall bladder

Chemical digestion takes place in the .... and is carried out by ....

[2]

4 Which statement best describes chemical digestion? Circle the correct letter.

  A breakdown of large insoluble molecules into small soluble molecules

  B breakdown of large soluble molecules into small insoluble molecules

  C joining of small insoluble molecules into large soluble molecules

  D joining of small soluble molecules into large insoluble molecules [1]

5 Explain why the food we eat needs to be broken down chemically.

SUPPLEMENT

....

.... [2]

6 Complete the table to show information on the different digestive enzymes.

SUPPLEMENT

Name of enzyme	Where it is produced	Action
	salivary glands and pancreas	breaks down starch into simple reducing sugars
Protease	stomach wall and pancreas	
Lipase		breaks down fats and oils into fatty acids and glycerol

[3]

> **TIP**
> Remember that enzyme names nearly always end in 'ase'.

7 The stomach contains hydrochloric acid in gastric juice.

SUPPLEMENT Describe **two** functions of the hydrochloric acid.

1 ....

2 .... [2]

B8 Transport in plants
Xylem and phloem

Student's Book pages 96–98 | Syllabus learning objectives B8.1.1–B8.1.2

1 a Complete the sentences about transport in plants using words from the list. Each word can be used once, more than once or not at all.

cellulose   leaves   petals   roots   seeds   water

Plants contain vessels that transport .... and dissolved substances around the plant. These vessels are found in the ...., stem and .... of the plant. [3]

The diagram shows a cross-section through a plant stem.

b Add the labels to the diagram by writing the name of the vessel in each box.

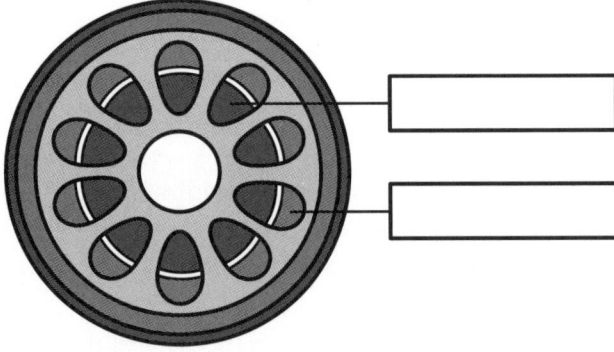

[2]

c State **two** substances transported by xylem vessels.

1 .... 2 .... [2]

d Phloem transports dissolved sugars, such as sucrose.

State **one other** substance transported by phloem.

.... [1]

2 Pests such as aphids feed from the xylem and phloem in the stems of wheat plants. Explain how this could reduce the growth of the wheat plant.

....

.... [2]

Water uptake

Student's Book pages 98–99 | Syllabus learning objectives B8.2.1–B8.2.2; **SUPPLEMENT** B8.2.3

1 **a** What is the main function of root hair cells? Circle the correct letter.

A to absorb amino acids

B to absorb soil particles

C to absorb sugar

D to absorb water and mineral ions [1]

b The diagram shows the movement of water across a root. State the names of the structures labelled **A**, **B** and **C**.

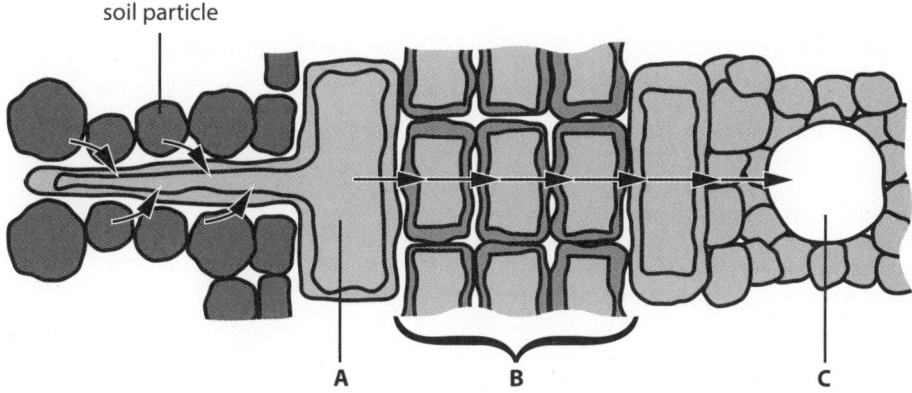

A .... B ....

C .... [3]

c Explain how the structure of a root hair cell is related to its function.

SUPPLEMENT

....

....

[2]

Transpiration

Student's Book pages 99–102 | Syllabus learning objectives B8.3.1; SUPPLEMENT B8.3.2

1 The diagram shows a magnified cross-section of the underside of a leaf.

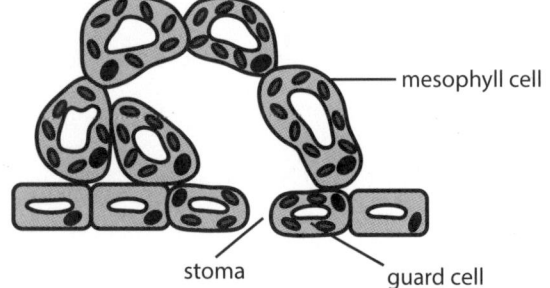

mesophyll cell

stoma

guard cell

TIP

Remember, stomata are tiny holes found in the epidermis of a leaf. Each separate hole is called a stoma.

a Draw an arrow onto the diagram to show the overall direction of movement of the water particles during transpiration. [1]

b Which statement best describes transpiration? Circle the correct letter.

A the absorption of water from the soil into the roots

B the loss of water vapour from the leaves

C the movement of dissolved sugars around the plant

D the movement of water through the plant [1]

2 A student uses a mass potometer as shown in the diagram to investigate the effect of changing air temperature on the rate of transpiration.

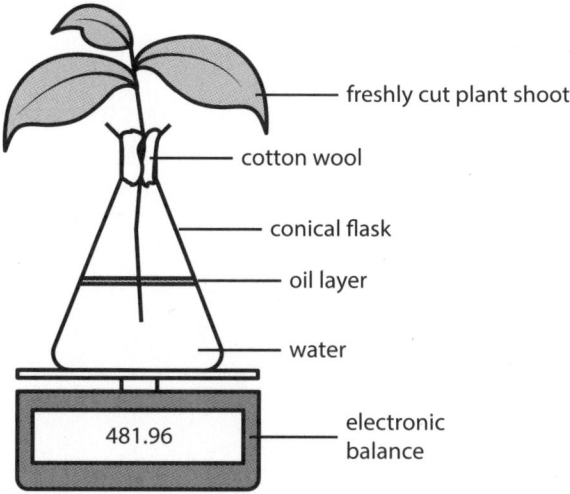

a State **two** control variables that the student would need to keep the same during the investigation.

1 ....

2 .... [2]

b Explain why oil is added to the conical flask.

.... [1]

c The student investigates the effects of changing the wind speed as well as the temperature.

Which set of conditions would produce the fastest decrease in mass?

Circle the correct letter.

	Temperature	Wind speed
A	high	high
B	high	low
C	low	high
D	low	low

[1]

B9 Transport in animals
Circulatory systems

Student's Book pages 118–119 | Syllabus learning objectives B9.1.1

1 Complete the sentences using words from the list. Each word can be used once, more than once or not at all.

| brain | cell | heart | lungs | one-way | two-way |

The circulatory system is made up of the .... and the blood vessels that transport blood to every .... in your body. Valves, found in some parts of the circulatory system, ensure a .... flow of blood. [3]

Heart

Student's Book pages 119–124 | Syllabus learning objectives B9.2.1–B9.2.4; **SUPPLEMENT** B9.2.5–B9.2.8

1 The diagram shows a heart.

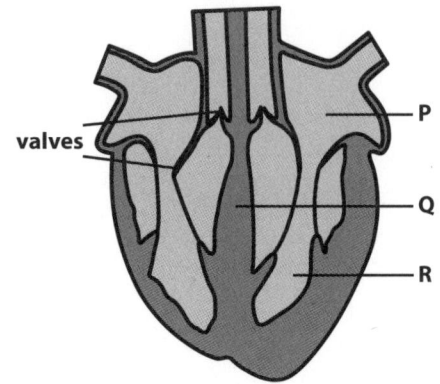

a) State the names of the parts of the heart labelled **P**, **Q** and **R**.

P ....

Q ....

R .... [3]

B9 Transport in animals | Heart

b Draw **four** arrows on the diagram to show the direction of blood flow into and out of the heart on the left side and the right side. [2]

c Describe the function of the valves in the heart.

SUPPLEMENT

.... [1]

d Which statement best describes arteries and veins? Circle the correct letter.

A Arteries always carry oxygenated blood, and veins always carry deoxygenated blood.

B Arteries are found on the left side of the heart, and veins are found on the right side of the heart.

C Arteries pump blood away from the heart, and veins return blood to the heart.

D Veins pump blood away from the heart, and arteries return blood to the heart. [1]

2 The table shows data on the thickness of the muscle walls found in the left and right ventricles of a sample of six adults aged 20–25 years old.

Adult	Left ventricle wall thickness (mm)	Right ventricle wall thickness (mm)
A	12.4	2.7
B	14.1	4.3
C	13.9	4.0
D	14.8	4.7
E	12.6	2.9
F	13.7	3.6

37

Calculate the mean thickness of the left and right ventricle walls of the heart.

Mean thickness of left ventricle wall: .... mm

Mean thickness of right ventricle wall: .... mm [2]

3 An electrocardiogram (ECG) is used by doctors to look at electrical waves in the heart as it beats. The line on the ECG graph below shows the electrical waves in the heart of a healthy person at rest.

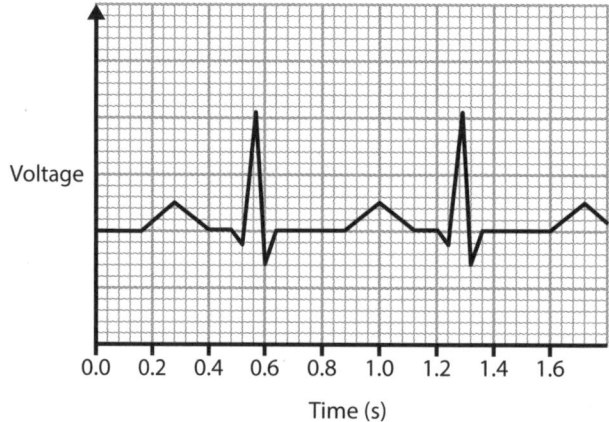

a Use the information in the graph to calculate the heart rate of the person.

Heart rate in beats per minute .... [2]

b Describe how the ECG would change during exercise.

.... [1]

c Suggest an alternative method to measure the activity of the heart at rest.

.... [1]

B9 Transport in animals | Blood vessels

4 **SUPPLEMENT** An athlete runs a 100 m sprint wearing a heart rate monitor. During the sprint, the athlete's heart rate changes from 45 beats per minute to 180 beats per minute.

a Explain why the heart rate increases so rapidly during the sprint.

....

....

.... [3]

b An athlete who exercises regularly could still have coronary heart disease.

State **three** risk factors that could increase an athlete's risk of getting coronary heart disease.

1 ....

2 ....

3 .... [3]

c State **two** ways an athlete could reduce their risk of coronary heart disease.

1 ....

2 .... [2]

Blood vessels

Student's Book pages 125–126 | Syllabus learning objectives B9.3.1; **SUPPLEMENT** B9.3.2

1 The diagram shows a magnified cross-section of an artery and a vein.

artery wall   vein wall

a Describe **one** difference between the artery wall and the vein wall.

....

.... [1]

b Explain how the structure of the artery wall is related to its function.

SUPPLEMENT

....

.... [2]

c Identify **one other** difference between the artery and vein that can be seen in the diagram.

.... [1]

> **TIP**
> When answering questions, try not to say 'it' unless it is absolutely clear what you are referring to. In this question, make sure it is clear whether you are writing about an artery or a vein.

d State **one** difference between the structure of arteries and veins that **cannot** be seen in the diagram.

.... [1]

e Capillaries are another type of blood vessel. State **one** way that the structure of capillaries is different from the structures of arteries and veins.

.... [1]

2 The graph shows the pressure of blood as it flows through the arteries, capillaries and veins of a person.

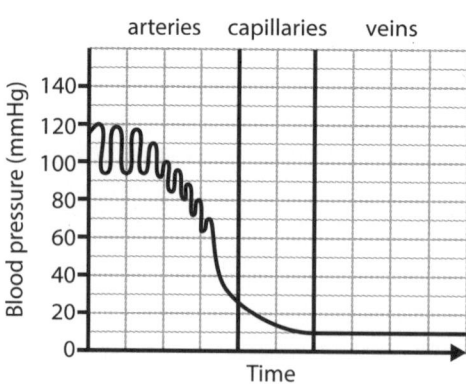

Calculate the difference between the maximum blood pressure in the arteries and the blood pressure in the veins.

.... mmHg [2]

Blood

Student's Book pages 127–129 | Syllabus learning objectives B9.4.1–B9.4.3; SUPPLEMENT B9.4.4

1 The table gives information about the components of blood and their functions.

a Complete the table by filling in the missing information.

Component of blood	Function
	transports dissolved substances in the blood
Red blood cells	
	protect against infection
Platelets	

[4]

b State **three** dissolved substances transported in the blood.

1 .... 2 ....

3 ....

[3]

2 This is a photomicrograph of human blood.

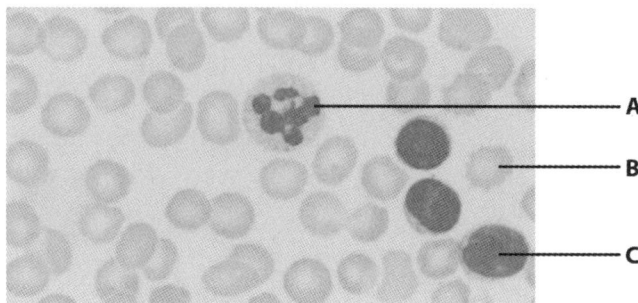

a Identify the cells labelled **A**, **B** and **C**.

A ....

B ....

C .... [3]

b Approximately 55% of the blood is plasma. An average human has 6500 cm³ of blood.

Calculate the volume of plasma in an average human body.

.... cm³ [1]

3 The table shows some blood test results for three people.

	Red blood cells (10^{12}/L)	White blood cells (10^9/L)	Platelets (10^9/L)
Normal range	5.0–7.0	10.5–25.5	100–450
Person A	6.1	8.7	365
Person B	5.4	12.8	201
Person C	8.1	14.5	437

a Calculate the difference in the platelet measurement between person **A** and person **B**.

.... [1]

b Regular training at high altitude increases the capacity of the blood to carry oxygen around the body.

State the letter of the person whose blood test result suggests they have been training at high altitude. .... [1]

c Platelets play an important role in clotting. State **two** functions of blood clotting.

SUPPLEMENT

1 ....

2 .... [2]

B10 Diseases and immunity
Diseases and immunity

Student's Book pages 136–142 | Syllabus learning objectives B10.1.1–B10.1.4;
SUPPLEMENT B10.1.5–B10.1.8

1 a Complete the sentences about disease using words from the list. Each word can be used once, more than once or not at all.

    contaminant   degenerative   host   transmissible   virus

Diseases that can be passed on from one organism to another by a pathogen are called .... diseases. An organism that is infected by a pathogen is called a .... [2]

b Which statement best describes a pathogen? Circle the correct letter.

A a microorganism that causes disease

B a poisonous substance

C a toxin produced by a microorganism

D an insect that causes disease [1]

c Circle **two** words that are possible pathogens.

    bacteria   cancer   mercury   toxins   viruses   [2]

2 a The human body has many barriers to help stop pathogens infecting the body. For example, the hairs inside the nose trap pathogens contained in air that is breathed in.

State **two other** examples of barriers that prevent pathogens causing infection in a human.

1 ....

2 .... [2]

b) Pathogens can be transferred directly or indirectly.

Complete the table by adding **one** tick (✓) to **each** row to show whether the description is an example of direct or indirect transmission.

Description of method of transmission	Direct	Indirect
Drinking contaminated water		
Contact with infected blood through sharing needles		
Being bitten by an infected animal		

[2]

c) Dysentery is an infectious disease that caused by the pathogen *Entamoeba histolytica*.

SUPPLEMENT

Dysentery causes diarrhoea and inflammation of the intestines. It is transmitted by eating food that is contaminated by the faeces of somebody who has dysentery, or by touching something that has *E. histolytica* on it, such as a toilet door handle.

Explain why personal hygiene is important in preventing the transmission of dysentery.

....

....

....

....

[2]

3 a) Describe how active immunity defends the body against disease.

SUPPLEMENT

....

....

[2]

b) State **two** ways a person can get active immunity to a disease.

1 ....

2 ....

[2]

45

B10 Diseases and immunity | Diseases and immunity

4 **SUPPLEMENT** There are many different types of virus. However, they all have the same simple structure. Describe the simple structure of a virus.

....

.... [2]

5 In the United Kingdom, the measles vaccination was introduced in 1968.

The table shows the number of cases of measles and the number of deaths from measles in the UK after the measles vaccination was introduced.

Year	Number of measles cases	Number of measles deaths
1970	307 408	42
1980	139 487	19
1990	13 302	10
2000	2378	4
2010	2235	1

a Calculate how the number of **measles cases** changed from 1970 to 2010.

.... [1]

b) Draw a line graph to show how the number of **measles deaths** changed from 1970 to 2010.

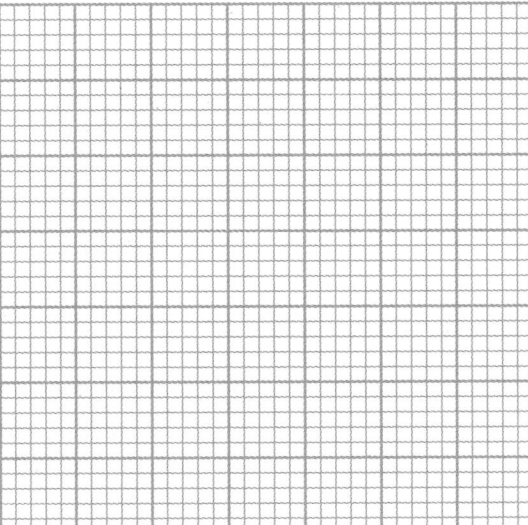

[4]

> **TIP**
>
> When drawing graphs, always use a pencil, **not** a pen. That way you can easily rub bits out and change them if necessary.

B11 Gas exchange in humans
Gas exchange in humans

Student's Book pages 148–153 | Syllabus learning objectives B11.1.1–B11.1.2; SUPPLEMENT B11.1.3

1 The diagram shows the human breathing system.

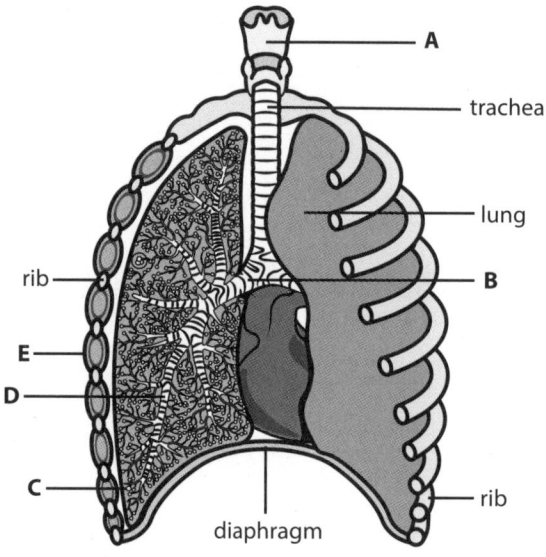

a Place letters in the boxes to show the names of the parts labelled **A**, **B**, **C** and **D**.

alveoli	
bronchiole	
bronchus	
larynx	

[3]

b State the name of the part labelled **E**.

....

[1]

2 Which feature is **not** found in human gas exchange surfaces? Circle the correct letter.

SUPPLEMENT

    A good blood supply

    B good ventilation with air

    C large surface area

    D thick surface

[1]

3 How does breathing change **after** physical activity? Circle the correct letter.

	Rate of breathing	Depth of breathing
A	decreases	decreases
B	decreases	increases
C	increases	decreases
D	increases	increases

[1]

B12 Respiration
Respiration

Student's Book pages 158–160 | Syllabus learning objectives B12.1.1–B12.1.3; **SUPPLEMENT** B12.1.4

1 One use of the energy released from respiration is to help maintain a constant body temperature.

State **two other** uses of energy in living organisms.

1 ....

2 .... [2]

2 a Complete the sentence about aerobic respiration using words from the list. Each word can be used once, more than once or not at all.

chemical   energy   light   nuclear   nutrient   small

The term aerobic respiration describes the ....

reactions in cells that use oxygen to break down ....

molecules to release .... [3]

b What is the word equation for aerobic respiration? (Circle) the correct letter.

A   carbon dioxide + glucose → water + oxygen

B   carbon dioxide + water → glucose + oxygen

C   glucose + oxygen → carbon dioxide + water

D   glucose + water → carbon dioxide + oxygen [1]

c Complete the balanced symbol equation for aerobic respiration.

SUPPLEMENT

.... + ... → .... + ... [2]

B13 Drugs
Drugs

Student's Book pages 164–166 | Syllabus learning objectives B13.1.1–B13.1.4; **SUPPLEMENT** B13.1.5

1 Which statement describes **all** drugs? Circle the correct letter.

  A substances that affect chemical reactions in the body

  B substances that can only be prescribed by a doctor

  C substances that kill bacteria in the body

  D substances that treat diseases [1]

2 What type of infections do antibiotics treat? Circle the correct letter.

  A bacterial infections   **C** bacterial and viral infections

  B viral infections   **D** neither bacterial nor viral infections [1]

3 Which statement describes antibiotic-resistant bacteria? Circle the correct letter.

  A Antibiotics are not effective against the bacteria.

  B Antibiotics are very effective against the bacteria.

  C The bacteria are immune to antibiotics.

  D The bacteria feed on antibiotics. [1]

4 Explain how doctors can limit the development of antibiotic-resistant bacteria.
SUPPLEMENT

....

.... [1]

B14 Reproduction
Sexual reproduction in plants

Student's Book pages 172–180 | Syllabus learning objectives B14.1.1–B14.1.5; SUPPLEMENT B14.1.6

1 Complete the sentence about pollination using terms from the list. Each term can be used once, more than once or not at all.

anthers   **ovules**   **petals**

pollen grains   **seeds**   **stigmas**

Pollination is the transfer of ....

from .... to .... [3]

2 The diagram shows the structure of an insect-pollinated flower.

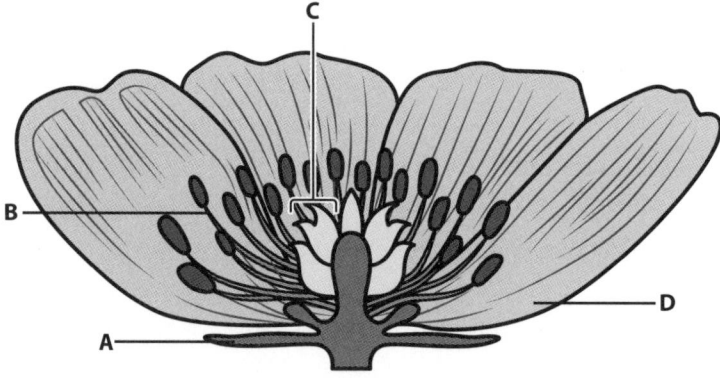

a Identify parts **A**, **B**, **C** and **D** and state their functions.

A: name ....

function ....

B: name ....

function ....

52

C: name ....

function ....

D: name ....

function .... [8]

b Describe the anthers and stigmas of a wind-pollinated flower.

SUPPLEMENT

anthers ....

....

stigmas ....

.... [2]

3 Which statement describes fertilisation in plants? Circle the correct letter.

A A pollen nucleus divides.

B A pollen nucleus fuses with a nucleus in an ovule.

C A pollen nucleus travels to an ovule.

D A pollen nucleus travels to a stigma. [1]

4 Place **three** ticks (✓) in the correct boxes to show which environmental conditions are needed for seeds to germinate.

carbon dioxide	
light	
oxygen	
suitable temperature	
water	

[3]

Sexual reproduction in humans

Student's Book pages 180–182 | Syllabus learning objectives B14.2.1–B14.2.4

1 The diagrams show the structures of the human female and male reproductive systems.

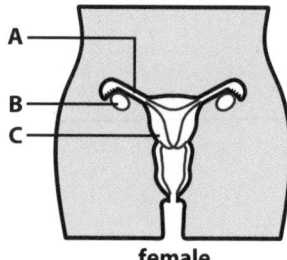
female

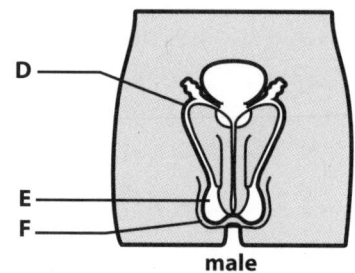
male

Identify parts **A**, **B**, **C**, **D**, **E** and **F** and state their functions.

A: name ....

function ....

B: name ....

function ....

C: name ....

function ....

D: name ....

function ....

E: name ....

function ....

F: name ....

function .... [12]

2 Complete the sentence about human reproduction using words from the list. Each word can be used once, more than once or not at all.

      divide   **egg**   **fuse**   **grow**   **ovary**

When a sperm fertilises an ...., their two nuclei

.... [2]

3 Which event does **not** occur during the menstrual cycle? Circle the correct letter.

  A egg implants in uterus lining

  B egg is released from an ovary

  C uterus lining breaks down

  D uterus lining thickens [1]

B15 Organisms and their environment

Energy flow

Student's Book page 193 | Syllabus learning objectives B15.1.1–B15.1.2

1 Complete the sentences about energy flow using words from the list. Each word can be used once, more than once or not at all.

chemical   **environment**   **kinetic**   **light**   **Moon**   **Sun**

Energy flows through food chains. The principal source of energy for food chains is .... energy from the .... Energy is transferred along food chains as .... energy in organisms. Energy in food chains is eventually transferred to the .... [4]

Food chains and food webs

Student's Book pages 194–198 | Syllabus learning objectives B15.2.1–B15.2.9; **SUPPLEMENT** B15.2.10

1 The table shows the meanings of some terms used when describing food chains and food webs.

Complete the table by choosing terms from the list. Each term can be used once, more than once or not at all.

carnivore   **consumer**   **decomposer**   **herbivore**

producer   **secondary**   **tertiary**

Term	Meaning
	an animal that gets its energy by eating other animals
	an animal that gets its energy by eating plants
	an organism that gets its energy by feeding on other organisms
	an organism that gets its energy from dead or waste organic material
	an organism that makes its own organic nutrients

[5]

2 This is a food chain.

grass → insects → mice → owls

a Identify which of the organisms is the producer.

.... [1]

b Identify which of the organisms is the secondary consumer.

.... [1]

c Add to the food chain to make a food web showing the following information:

- Insects are also eaten by small birds called sparrows.

- Sparrows are eaten by owls.

- Foxes eat mice and sparrows. [3]

d  The number of sparrows in the UK is decreasing. This may be because of environmental changes caused by humans.

Use the food chain and your additions to it to explain the effect a decrease in the number of sparrows may have on the number of mice.

....

.... [1]

Carbon cycle

Student's Book pages 199–200 | Syllabus learning objective B15.3.1

1 The diagram shows part of the carbon cycle.

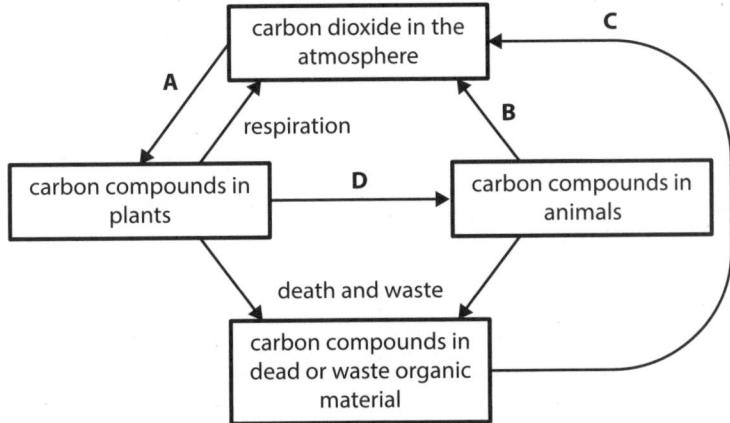

The boxes on the left below show letters from the diagram of the carbon cycle. The boxes on the right show some processes in the carbon cycle.

Draw lines to link each letter to the process it represents. Draw **four** lines.

Letter	Process
A	combustion
B	decomposition
C	feeding
D	photosynthesis
	respiration

[4]

B16 Human influences on ecosystems
Habitat destruction

Student's Book pages 206–210 | Syllabus learning objectives B16.1.1–B16.1.4;
SUPPLEMENT B16.1.5

1 Which statement describes an ecosystem? Circle the correct letter.

　A　a community of organisms

　B　a community of organisms and their environment, interacting together

　C　a species and its environment, interacting together

　D　the environment where a community of organisms lives [1]

2 Which statement describes biodiversity? Circle the correct letter.

　A　the number of different communities that live in an area

　B　the number of different species that have ever lived in an area

　C　the number of different species that live in an area

　D　the number of individuals of one species that live in an area [1]

3 Deforestation is one example of habitat destruction.

　a　Describe **two** reasons for habitat destruction.

　1 ....

　2 .... [2]

b One effect of deforestation is an increase in carbon dioxide in the atmosphere. State **two other** undesirable effects of deforestation.

1 ....

....

2 ....

.... [2]

c Explain how deforestation leads to an increase of carbon dioxide in the atmosphere.

SUPPLEMENT

....

....

....

.... [4]

Conservation

Student's Book pages 210–212 | Syllabus learning objectives B16.2.1; SUPPLEMENT B16.2.2

1 What term describes a species that is at risk of dying out? Circle the correct letter.

    A conserved

    B endangered

    C extinct

    D protected [1]

B16 Human influences on ecosystems | Conservation

2 **SUPPLEMENT** A rare species of tree only grows in an area at risk from deforestation.

Describe **two** ways the species can be protected from dying out.

1 ....

....

2 ....

.... [2]

Chemistry

C1 States of matter — 65
Solids, liquids and gases — 65

C2 Atoms, elements and compounds — 67
Elements, compounds and mixtures — 67
Atomic structure and the Periodic Table — 69
Ions and ionic bonds — 71
Simple molecules and covalent bonds — 75

C3 Stoichiometry — 77
Formulas — 77

C4 Electrochemistry — 80
Electrolysis — 80

C5 Chemical energetics — 82
Exothermic and endothermic reactions — 82

C6 Chemical reactions — 86
Physical and chemical changes; Rates of reaction — 86
Redox — 90

C7 Acids, bases and salts — 92
The characteristic properties of acids and bases; Oxides — 92
Preparation of salts — 94

C8 The Periodic Table — 96
Arrangement of elements — 96
Group I properties — 98
Group VII properties — 99
Transition elements; Noble gases — 101

C9 Metals — 103
Properties of metals — 103
Uses of metals; Alloys and their properties — 105
Reactivity series; Corrosion of metals — 107
Extraction of metals — 109

C10 Chemistry of the environment — 112
Water — 112
Air quality and climate — 113

C11 Organic chemistry — 116
Fuels — 116
Alkanes — 118
Alkenes — 119
Polymers — 121

C12 Experimental techniques and chemical analysis — 122
Experimental design — 122
Chromatography — 124
Separation and purification — 127
Identification of ions and gases — 129

C1 States of matter
Solids, liquids and gases

Student's Book pages 222–227 | Syllabus learning objectives C1.1.1–C1.1.4; SUPPLEMENT C1.1.5

1 How are the arrangement and movement of particles in a solid different from those in a gas?

....

.... [2]

2 How do the arrangement and movement of particles in a liquid differ from those in a solid?

....

.... [2]

3 What apparatus would you use to measure the volume of a liquid?

.... [1]

4 Look at the particle diagrams below.

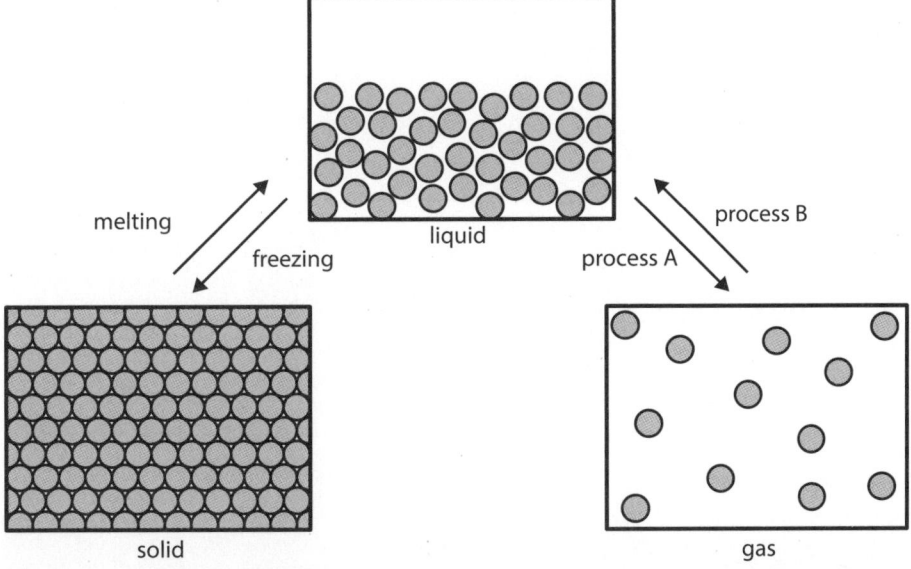

Label the two processes marked on the diagram.

Process A: .... [1]

Process B: .... [1]

5 Look at the diagrams below.

a What change in temperature would make balloon A look more like balloon B?

.... [1]

b What change in pressure would make balloon B look more like balloon A?

.... [1]

> **TIP**
> Check your understanding of the kinetic particle theory in the Student's Book.

6 Use kinetic particle theory to explain the changes that occur when a solid turns into a liquid.

SUPPLEMENT

....

.... [2]

C2 Atoms, elements and compounds
Elements, compounds and mixtures

Student's Book page 232 | Syllabus learning objective C2.1.1

> **TIP**
> The Periodic Table only includes elements.

1 What is the difference between an element and a compound?

....

.... [2]

2 What is the difference between a compound and a mixture?

....

.... [2]

3 Complete the following table by selecting the correct description.

Substance	Element	Compound	Mixture
Sea water			
Sodium chloride			
Copper(II) sulfate solution			
Iron			
Diamond			

Substance	Element	Compound	Mixture
Distilled/pure water			
Oxygen			
Carbon dioxide			
Air			

[9]

4 Look at the diagram below. Each of the spheres represents an element.

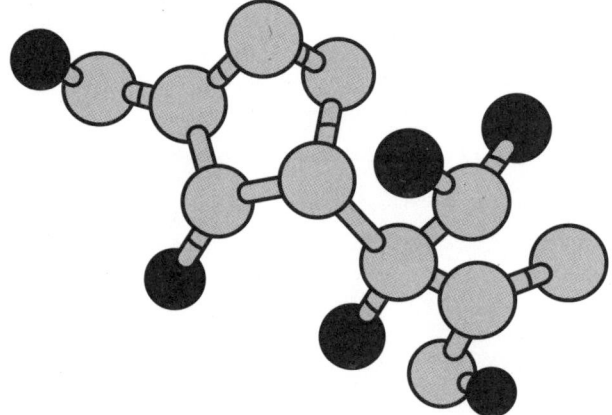

a Does the diagram represent an element, a compound or a mixture?

....

[1]

b Explain your answer.

....

....

[2]

Atomic structure and the Periodic Table

Student's Book pages 233–238 | Syllabus learning objectives C2.2.1–C2.2.6

1 State whether the following statements are **True** or **False**.

Statement	True or False
Protons are found in the nucleus.	
In an atom, the numbers of protons and neutrons are always the same.	
Electrons are arranged in shells around the nucleus.	
An electron has a relative mass of 1	
A proton has a relative charge of +1	
The nucleon number gives the number of protons and neutrons in the nucleus.	
An element with an atomic number of 11 has three electron shells.	

[7]

TIP

When answering questions about atomic structure, make sure to have a copy of the Periodic Table to hand (included on page 215 of this Workbook).

2 Complete the following table.

Atom	Atomic number	Nucleon number	Number of neutrons	Number of electrons	Electron arrangement
$^{7}_{3}Li$					
$^{19}_{9}F$					
$^{28}_{14}Si$					
$^{31}_{15}P$					
$^{39}_{19}K$					

[5]

3 Look at the Periodic Table on page 215 and then answer questions **a–e**.

a Which group is nitrogen in?

.... [1]

b Which group is calcium in?

.... [1]

c Sodium is in the third Period. How many electron shells does it have?

.... [1]

d Oxygen is in Group VI. How many electrons does it have in its outer electron shell?

.... [1]

e What do the final electron shells of all the noble gases have in common?

.... [1]

4 The drawing below shows an atom diagram.

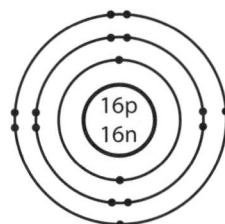

Taking information from the Periodic Table on page 215, draw atom diagrams of the following elements:

a Aluminium [2]

b Magnesium [2]

c Neon [2]

Ions and ionic bonds

Student's Book pages 242–247 | Syllabus learning objectives C2.3.1–C2.3.4; SUPPLEMENT C2.3.5–C2.3.6

1 Chlorine forms the chloride ion Cl^-.

a Has the chlorine atom lost or gained an electron?

[1]

b Why is only one electron lost or gained?

[1]

c Is the chloride ion an anion or a cation?

[1]

C2 Atoms, elements and compounds | Ions and ionic bonds

2 Sodium chloride is a solid at room temperature.

a What holds the solid structure together?

.... [1]

b Why does solid sodium chloride not conduct an electric current?

.... [1]

c Would you expect sodium chloride to have a high or low melting point?

.... [1]

d Describe what you would expect to happen if sodium chloride is added to water.

.... [1]

3 Use the atomic numbers in the Periodic Table on page 215 to help you complete the table below.

Element	Electron arrangement of the atom	Electron arrangement of the ion	Charge on the ion
Sodium			
Fluorine			
Potassium			

[9]

4 Draw dot-and-cross diagrams to show the formation of the ionic bonds between the following elements. In each case, write the formulas of the ions formed.

> **TIP**
> You will need to use the atomic numbers of the elements as shown in the Periodic Table on page 215.

a Potassium fluoride

[2]

b Lithium chloride

[2]

5 The following diagram represents the structure of sodium chloride.

SUPPLEMENT

● chloride ion   ○ sodium ion

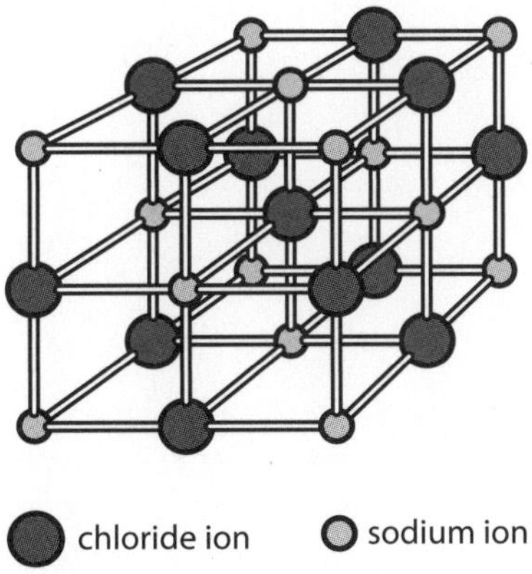

What does the diagram show about the structure of sodium chloride?

....

....

.... [3]

6 Draw dot-and-cross diagrams to show the ionic bonding in the following compounds. In each case, write the formulas of the ions formed.

SUPPLEMENT

> **TIP**
> Use the same approach as in Chapter 2 of the Student's Book.

a Calcium oxide (proton numbers Ca = 20; O = 8)

[2]

b Aluminium chloride (proton numbers Al = 13; Cl = 17)

[2]

Simple molecules and covalent bonds

Student's Book pages 250–255 | Syllabus learning objectives C2.4.1–C2.4.3; **SUPPLEMENT** C2.4.4

1 State how a covalent bond is formed.

.... [1]

2 Use a dot-and-cross diagram to show how the covalent bond is formed in chlorine, Cl_2. The proton number of chlorine is 17.

[2]

3

a Use a dot-and-cross diagram to show the covalent bonds in ammonia, NH_3. The proton number of hydrogen is 1 and the proton number of nitrogen is 7.

[2]

C2 Atoms, elements and compounds | Simple molecules and covalent bonds

b) Will ammonia conduct electricity? Give a reason for your answer.

....

.... [2]

c) Would you expect ammonia to have a low or high boiling point? Give a reason for your answer.

....

.... [2]

4 SUPPLEMENT Draw a dot-and-cross diagram to show the electron configuration of ethene, C_2H_4. The proton number of hydrogen is 1 and the proton number of carbon is 6.

[3]

C3 Stoichiometry
Formulas

Student's Book pages 260–268 | Syllabus learning objectives C3.1.1–C3.1.5;
SUPPLEMENT C3.1.6– C3.1.7

1 Deduce the formulas of the following compounds:

> **TIP**
> Use the Periodic Table on page 215 to identify the combining powers of the elements in the compound.

a Sodium oxide

[1]

b Hydrogen sulfide

[1]

c Aluminium chloride

[1]

2 Deduce the chemical formulas of the following compounds:

a Copper(II) sulfate

[1]

b Calcium carbonate

[1]

c Sodium hydroxide

[1]

C3 Stoichiometry | Formulas

3 Look at the diagrammatic representation of ethanol. Deduce the molecular formula of ethanol.

$$H-\underset{\underset{H}{|}}{\overset{\overset{H}{|}}{C}}-\underset{\underset{H}{|}}{\overset{\overset{H}{|}}{C}}-OH$$

.... [1]

4 Magnesium burns in oxygen to form a white powder which is magnesium oxide.

Write a word equation for the reaction, including state symbols.

.... [2]

5 Methane burns in oxygen to form carbon dioxide and water.

SUPPLEMENT

a Write a word equation for the reaction, including state symbols.

.... [2]

b Write a symbol equation for the reaction, including state symbols.

....

.... [2]

6 The following symbol equation represents a chemical reaction.

SUPPLEMENT

$$CuSO_4 \text{ (aq)} + 2NaOH \text{ (aq)} \rightarrow Cu(OH)_2 \text{ (s)} + Na_2SO_4 \text{ (aq)}$$

a Name the **two** reactants.

1 .... [1]

2 .... [1]

b Explain what the '2' in front of 'NaOH' tells us.

.... [1]

c Using your knowledge of state symbols, describe what a scientist carrying out this reaction would observe.

....

.... [2]

> **TIP**
> Refer to the Student's Book on how to write ionic equations.

7 Balance the following symbol equations:

a ___$Cu(NO_3)_2$(s) → ___ CuO(s) + ___ NO_2(g) + ___ O_2(g) [1]

b ___$CaCO_3$(s) + ___HCl(aq) → ___$CaCl_2$(aq) + ___CO_2(g) + ___H_2O(l) [1]

8 Balance the following ionic equations:

SUPPLEMENT

a ___Fe^{3+}(aq) + ___OH^-(aq) → ___$Fe(OH)_3$(s) [1]

b ___H^+(aq) + ___CO_3^{2-}(aq) → ___H_2O(l) + ___CO_2(g) [1]

9 The sodium ion has a single positive charge and the oxide ion has a double negative charge.

SUPPLEMENT

a Write the symbols for these ions.

.... [1]

b Deduce the formula of sodium oxide.

.... [1]

C4 Electrochemistry
Electrolysis

Student's Book pages 278–282 | Syllabus learning objectives C4.1.1–C4.1.3;
SUPPLEMENT C4.1.4–C4.1.5

1 Use this diagram of some electrolysis apparatus to answer the questions that follow.

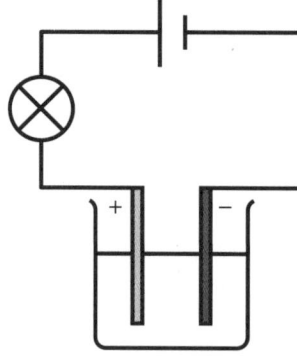

a What name is given to the two shaded rods in the diagram?

.... [1]

b What inert materials can be used for these rods?

.... [1]

c What name is given to the rod marked with a plus sign?

.... [1]

d What name is given to the rod marked with a minus sign?

.... [1]

e The bulb in the circuit lights up. What name is given to the type of liquid in the beaker?

.... [1]

2 What products are formed in the electrolysis of concentrated aqueous sodium chloride solution?

a At the anode:

.... [1]

b At the cathode:

.... [1]

3 What products will be formed in the electrolysis of molten lead(II) bromide?

a At the anode:

.... [1]

b At the cathode:

.... [1]

4 Using the same apparatus as shown in Question 1, an electric current is passed through molten sodium chloride.
SUPPLEMENT

a What product will be formed at the anode?

.... [1]

b What product will be formed at the cathode?

.... [1]

c If the substance in the beaker is solid sodium chloride, explain why no electrolysis will take place.

....

.... [2]

C5 Chemical energetics
Exothermic and endothermic reactions

Student's Book pages 288–295 | Syllabus learning objectives C5.1.1–C5.1.2; **SUPPLEMENT** C5.1.3–C5.1.6

1 Which of the following statements is correct? Circle the correct letter.

   A An exothermic reaction absorbs energy from the surroundings.

   B An endothermic reaction releases thermal energy.

   C An exothermic reaction causes an increase in the temperature of the surroundings.

   D An endothermic reaction does not absorb or release thermal energy. [1]

2 The apparatus shown in the diagram was used by three groups of students (groups A, B and C) to compare the thermal energy produced by burning ethanol.

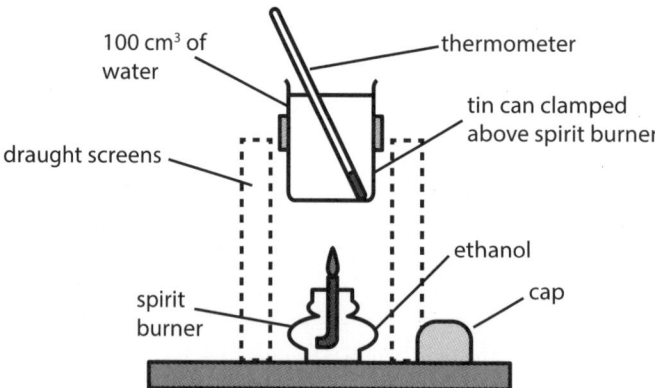

The results produced by the three groups (A–C) are shown in the table.

Group	Temperature rise of the water (°C)	Mass of spirit burner and ethanol before burning (g)	Mass of spirit burner and ethanol after burning (g)	Mass of ethanol burnt (g)	Temperature rise produced by burning 1 g of ethanol (°C/g)
A	34	52.2	51.4		
B	32	50.6	49.9		
C	36	51.4	50.5		

a Complete the second last column in the table. [3]

b Complete the final column in the table. Give your answers to 3 significant figures. [3]

c Which group's results show the highest thermal energy produced when burning ethanol?

.... [1]

> **TIP**
> Where you are asked to identify differences or errors, make sure you answer in terms of the actual experiment the question is referring to.

d Why was the water being heated in a tin can rather than a glass beaker? [2]

....

....

e Suggest some reasons why the temperature rises per gram of ethanol were not the same for each group.

....

.... [2]

> **TIP**
> Where a question about a particular chemical term has 2 marks, you must try to give an answer that has at least two pieces of information.

C5 Chemical energetics | Exothermic and endothermic reactions

3 Look at the diagram of the reaction pathway and then answer the questions that follow.

SUPPLEMENT

reaction pathway

a Label the vertical axis.

.... [1]

b Label each of the two horizontal lines linked by the downwards arrow.

.... [1]

c What type of reaction does this reaction pathway diagram illustrate?

.... [1]

4 Explain what is meant by the term activation energy, E_a.

SUPPLEMENT

....

.... [2]

5 A reaction takes place with an activation energy E_a. The overall energy change in the reaction is E_r, where this energy is transferred from the reaction to the surroundings.

a Draw and label a reaction pathway diagram for this reaction in the space below.

[4]

b State whether the reaction is exothermic or endothermic.

[1]

6 Methane (CH_4) burns in oxygen to form carbon dioxide and water.

a Write down a fully balanced equation for the reaction.

[2]

b In the first stage of the reaction which bonds must be broken?

[2]

c Is bond breaking an exothermic or endothermic process?

[1]

d The next stage of the reaction is bond making. Is this an exothermic or endothermic process?

[1]

C6 Chemical reactions
Physical and chemical changes; Rates of reaction

Student's Book pages 302–312 | Syllabus learning objectives: C6.1.1; C6.2.1–C6.2.4; SUPPLEMENT C6.2.5–C6.2.6

1 Which **one** of the following experiments involves a chemical change? Circle the correct letter.

    A   adding a spatula measure of sodium chloride to a beaker of distilled water

    B   mixing gas jars of carbon dioxide and oxygen

    C   putting a graphite rod into a beaker of warm water

    D   adding magnesium ribbon to a beaker of dilute hydrochloric acid

[1]

2 State whether the following changes are physical or chemical

a Boiling a kettle of water

[1]

b Lighting a candle for a birthday cake

[1]

c Using petrol as the fuel in a car

[1]

d Adding sugar to a cup of coffee

[1]

C6 Chemical reactions | Physical and chemical changes; Rates of reaction

3 When copper(II) sulfate crystals are added to water in a conical flask, the water turns blue. Is this a physical or chemical change? Give a reason for your answer.

....

.... [2]

4 Magnesium ribbon reacts with dilute sulfuric acid, producing magnesium sulfate solution and hydrogen gas. Describe an experiment that you could use to work out the effect of temperature on the rate of reaction. You have to use three different temperatures: 30 °C, 35 °C and 40 °C.

TIP
When measuring the rate of a reaction, you must be able to measure the time taken.

a In the space below, draw and label the apparatus you would use in this experiment.

[3]

87

b In the space below, draw a table suitable to record your results.

[3]

5 An experiment is used to compare the reactions of magnesium powder and magnesium ribbon with an excess amount of dilute hydrochloric acid.

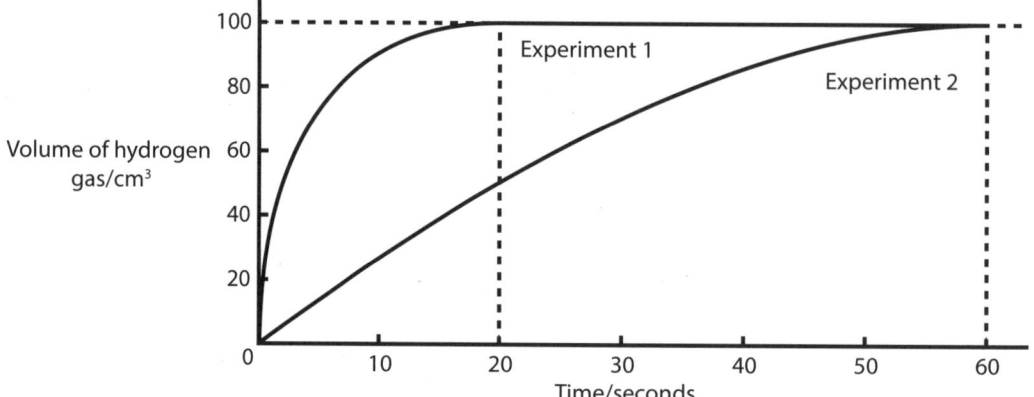

a In which experiment was the magnesium ribbon? Explain your reasoning.

....

.... [2]

b Why is the rate of producing hydrogen in Experiment 1 higher in the first 5 seconds than after 10 seconds?

....

.... [2]

c Why do both graphs level out?

.... [1]

d Suggest a reason why both graphs level out at 100 cm³ of hydrogen.

.... [1]

6 Give **two** important characteristics of a catalyst.

1 .... [1]

2 .... [1]

7 Using collision theory, describe and explain the effect on the rate of reaction of increasing the temperature in the reaction between marble chips and dilute hydrochloric acid.

SUPPLEMENT

....

....

.... [3]

8 The following apparatus is used to measure the effect of temperature on the volume of gas produced when magnesium is added to dilute hydrochloric acid.

You have been asked to compare the rate of reaction at four different temperatures, using the apparatus four times. State **two** conditions that must not be changed in the four experiments.

1 ....

2 .... [2]

9 State how a catalyst increases the rate of a reaction.

SUPPLEMENT

.... [1]

10 In the reaction between magnesium and hydrochloric acid, use collision theory to explain why increasing the concentration of the acid will increase the rate of reaction.

SUPPLEMENT

....

....

.... [3]

Redox

Student's Book pages 316–318 | Syllabus learning objectives C6.3.1–C6.3.4

1 What is a redox reaction?

.... [1]

2 In chromium chloride, chromium has an oxidation number of 3.

a Show how the oxidation number of chromium can be included when naming the compound.

.... [1]

b What is the formula of the chromium chloride?

.... [1]

3 Define the following in terms of oxygen:

a Oxidation

.... [1]

b Reduction

.... [1]

4 Complete the following table by adding a tick (✓) to show if each reaction is a redox reaction.

Reaction	Is this a redox reaction?
$MgCO_3(s) \rightarrow MgO(s) + CO_2(g)$	
$2Mg(s) + O_2(g) \rightarrow 2MgO(s)$	
$MgO(s) + H_2(g) \rightarrow Mg(s) + H_2O(l)$	

[3]

5 Look at the following equation and then answer questions **a**–**b**.

$$Zn(s) + CuO(s) \rightarrow ZnO(s) + Cu(s)$$

a What has been oxidised in this reaction?

.... [1]

b What has been reduced in this reaction?

.... [1]

C7 Acids, bases and salts
The characteristic properties of acids and bases; Oxides

Student's Book pages 322–326 | Syllabus learning objectives C7.1.1–C7.1.6; C7.2.1

1 Complete the following table.

Solution	Colour of litmus	Colour of methyl orange
Sodium hydroxide		
Sulfuric acid		

[4]

2 Complete the following table about the use of universal indicator paper.

> **TIP**
> You need to be very familiar with the pH scale.

Type of solution	pH value	Colour with universal indicator
Neutral		
Strongly acidic		
Weakly alkaline		
Weakly acidic		

[8]

3 Acids react with metals such as magnesium. Write a balanced equation, including state symbols, for the reaction between magnesium and dilute hydrochloric acid.

.... [2]

4 Nitric acid reacts with carbonates such as calcium carbonate.

a Write a word equation for this reaction.

.... [1]

b Write a balanced equation, including state symbols, for the reaction.

.... [2]

> **TIP**
> You need to ensure you can write balanced equations for the reactions you need to understand.

5 State what property of a base means that it is an alkali.

.... [1]

6 Which of the following is **not** a basic oxide? Circle the correct letter.

    A   zinc oxide

    B   aluminium oxide

    C   sodium oxide

    D   sulfur dioxide

[1]

Preparation of salts

Student's Book pages 327–332 | Syllabus learning objectives C7.3.1; **SUPPLEMENT** C7.3.2

1

a Explain the meaning of the term soluble.

[1]

b Explain the meaning of the term insoluble.

[1]

2 Zinc sulfate crystals can be prepared using the reaction between solid zinc carbonate and dilute sulfuric acid.

a Write a word equation for this reaction.

[1]

b Write a fully balanced equation for this reaction.

[2]

> **TIP**
> In the preparation of a soluble salt, there are many different stages, each using different apparatus.

c Describe a method for preparing zinc sulfate crystals, a soluble salt, in the laboratory. In your method, be organised in stages. You should include:

　i which of the two reactants should be in excess

　ii the apparatus needed at each stage in the preparation and what happens at each stage of the preparation.

....

....

....

....

....

.... [5]

3 **SUPPLEMENT** Copper(II) carbonate can be prepared using copper(II) nitrate solution and sodium carbonate solution. The three stages of the preparation are summarised below:

Stage 1. 30 cm^3 of copper(II) nitrate solution is mixed with 30 cm^3 of sodium carbonate solution. The copper(II) carbonate forms as a precipitate.

Stage 2. The copper(II) carbonate is separated from the solutions.

Stage 3. The solid copper(II) carbonate is dried.

a What apparatus is needed for **Stage 1**?

.... [1]

b What is a precipitate?

.... [1]

c Sketch and label the apparatus you would use for **Stage 2**.

[2]

C8 The Periodic Table
Arrangement of elements

Student's Book pages 344–347 | Syllabus learning objectives C8.1.1–C8.1.2;
SUPPLEMENT C8.1.3

1 Below is a representation of the Periodic Table with some of the elements marked with letters.

TIP

You do not need to learn the arrangement of the elements in the Periodic Table – a Periodic Table will be provided in your exams.

a Which element is in Group VII?

.... [1]

b Which elements are in the same group?

.... [1]

c Which elements are in the same period?

.... [1]

d Which elements are non-metals?

.... [1]

e Which element will have the greatest proton number/atomic number?

.... [1]

2 The following table shows some physical properties for the first three elements in Group I of the Periodic Table.

SUPPLEMENT

> **TIP**
> It may help you to know the density of iron is 7.86 g/cm^3.

Element	Melting point (°C)	Boiling point (°C)	Density g/cm^3
Lithium	180	1342	0.53
Sodium	98	883	0.97
Potassium	64	759	0.86

What trends can you identify in the melting point, boiling point and density information from the table?

....

....

.... [3]

Group I properties

Student's Book pages 350–354 | Syllabus learning objective C8.2.1; **SUPPLEMENT** C8.2.2

1 Which of the following statements about sodium is **untrue**? Circle the correct letter.

    A Sodium reacts with water to form sodium hydroxide, an alkali.

    B Sodium is stored under oil.

    C Sodium is a soft metal which is easily cut.

    D Sodium has the highest melting point of the elements in Group I. [1]

2 Which of the following statements is correct? Circle the correct letter.

    A The Group I elements are unreactive metals.

    B Potassium is the most reactive of the metals in the group.

    C Potassium has a higher density than lithium.

    D The Group I elements have high melting points. [1]

> **TIP**
> You may need to revise atomic structure from an earlier topic.

3 Potassium has a proton number/atomic number of 19.

What is the electron arrangement in potassium?

[1]

4 Sodium is added to some distilled water in a large beaker. State **two** observations you would expect to see.

1 ....

2 .... [2]

5 Rubidium is the element below potassium in Group I of the Periodic Table.

SUPPLEMENT

a Use your knowledge about lithium, sodium and potassium to predict whether rubidium is a metal or non-metal.

.... [1]

b How would you expect the reactivity of rubidium with water to compare to that of potassium?

.... [1]

Group VII properties

Student's Book pages 357–361 | Syllabus learning objectives C8.3.1–C8.3.2; SUPPLEMENT C8.3.3–C8.3.4

1 Explain the meaning of the term diatomic.

.... [1]

2 Iodine and bromine are both elements in Group VII.

a Describe the appearance of iodine and bromine at room temperature and pressure.

....

.... [2]

99

b) How many electrons are in the outer shell of iodine and bromine atoms?

.... [1]

c) Compare the densities and reactivities of iodine and bromine.

....

.... [2]

3 Fluorine is in Group VII.

SUPPLEMENT

a) Use your knowledge about chlorine, bromine and iodine to predict whether fluorine is a solid, liquid or gas at room temperature and pressure.

.... [1]

b) How would you expect the reactivity of fluorine to compare to that of chlorine?

.... [1]

c) When fluorine forms compounds, what ion does it form?

.... [1]

d) Use your knowledge of atomic structure to explain why an element in Group VII will form this ion.

....

.... [2]

4 When chlorine is bubbled into a solution of sodium bromide, an orange/brown solution is formed. This is an example of a displacement reaction.

SUPPLEMENT

a) What substance is responsible for the formation of the orange/brown solution?

.... [1]

b State what a halogen displacement reaction is.

....

.... [2]

c Write a word equation for the reaction between chlorine gas and sodium bromide solution.

.... [1]

d Write a balanced equation for the reaction between chlorine gas and sodium bromide solution.

.... [2]

Transition elements; Noble gases

Student's Book pages 364–366 | Syllabus learning objectives C8.4.1; C8.5.1

1 Which of the following is **not** a typical characteristic of transition elements? Circle the correct letter.

  A They form compounds with colours such as blue, green or brown.

  B They react vigorously when in contact with water.

  C They act as catalysts in industrial processes.

  D They are often used in construction. [1]

2 Copper is a transition metal whereas sodium is an alkali metal. Describe **three** differences in physical properties between these metals.

1 ....

2 ....

3 .... [3]

3 Helium is a monatomic gas. Explain what this means.

.... [1]

> **TIP**
> In your exams, you will have a Periodic Table (included on page 215) to use when answering these questions.

4 Neon is a very unreactive element.

a What is the proton number/atomic number of neon?

.... [1]

b How are the electrons arranged in an atom of neon?

.... [1]

c Use the electronic configuration to explain why neon is so unreactive.

....

.... [2]

C9 Metals
Properties of metals

Student's Book pages 372–374 | Syllabus learning objectives C9.1.1–C9.1.2

1 Which of the following is **not** a common/general physical property of a metal? Circle the correct letter.

    A   low melting point

    B   good conductor of electricity

    C   malleable

    D   good conductor of heat   [1]

2 A student sets up some apparatus to test the electrical conductivity of an element. Draw and label the apparatus the student could have used.

[3]

3 Metals are often ductile. Explain what this term means.

[1]

C9 Metals | Properties of metals

4 For each of the following reactions (**a–b**), write a word equation and a balanced symbol equation.

> **TIP**
> When writing equations, use the Periodic Table (included on page 215) to find the group the element is in and therefore its combining power/electronic charge. With the transition metals, look for the roman numerals in the name of any of its compounds – if these are not given, then (II) is a good guess!

a Magnesium reacting with steam.

   i Word equation:

.... [1]

   ii Symbol equation:

.... [2]

b Zinc reacting with dilute hydrochloric acid.

   i Word equation:

.... [1]

   ii Symbol equation:

.... [2]

5 Draw and label the apparatus that can be used in the reaction between magnesium and dilute sulfuric acid to collect the gas produced.

[3]

Uses of metals; Alloys and their properties

Student's Book pages 374–377 | Syllabus learning objectives C9.2.1; C9.3.1–C9.3.4; SUPPLEMENT C9.3.5

1 Aluminium is used in the manufacture of aircraft, usually in the form of an alloy.

a Define the term alloy.

.... [1]

b Give a reason why aluminium is used in the manufacture of aircraft.

.... [1]

c Why is the aluminium used in aircraft manufacture in the form of an alloy?

.... [1]

2 Give **two** reasons why copper is used in electrical wiring.

1 ....

2 .... [2]

3 Which **one** of the following statements best explains the use of aluminium in metal cans for storing food? Circle the correct letter.

   A Aluminium has a high melting point.

   B Aluminium resists oxidation and corrosion generally.

   C Aluminium has low density.

   D Aluminium conducts electricity. [1]

105

C9 Metals | Uses of metals; Alloys and their properties

4 Give **two** reasons why stainless steel is often used to make kitchen knives.

1 ....

2 .... [2]

5 Is an alloy a compound or a mixture? Explain your answer.

....

.... [2]

6 a **SUPPLEMENT** Sketch diagrams to show the difference in the arrangement of the atoms in aluminium and in an alloy of aluminium.

[2]

b Use the diagrams to explain why an aluminium alloy is stronger than the pure metal.

....

....

.... [2]

Reactivity series; Corrosion of metals

Student's Book pages 377–380 | Syllabus learning objectives C9.4.1–C9.4.3; C9.5.1–C9.5.3

1 The reactivity series of metals often includes the non-metal hydrogen. Suggest a reason for this.

[2]

2 The following table shows the results of a series of experiments performed with elements listed in the reactivity series.

Element	Reaction with cold water?	Reaction with steam?	Reaction with dilute hydrochloric acid?
W	No	No	No
Y	Yes	Yes	Yes
Z	No	Yes	Yes

a Arrange the elements (W–Z) in order of reactivity starting with the most reactive.

[1]

b Complete the following table by suggesting a possible name for each of the elements. [3]

Element	Possible name
W	
Y	
Z	

3

a Name a metal which, when added to cold water, floats on the surface and catches fire.

.... [1]

b Write a word equation for the reaction between this metal and water.

.... [1]

c Write a balanced equation for the reaction between this metal and water.

.... [2]

4 The reactivity series is made up mostly of metals. Explain why it is common to include carbon in the reactivity series.

....

.... [2]

5 In the right conditions iron will corrode to form rust.

Describe the conditions required for iron to corrode.

.... [2]

6 Which of the following methods will **not** prevent the rusting of iron? Circle the correct letter.

    A coating the iron with plastic

    B adding carbon to the iron

    C painting the iron

    D greasing the iron [1]

7 What are barrier methods and how do they prevent rusting?

....

.... [2]

Extraction of metals

Student's Book pages 380–384 | Syllabus learning objectives C9.6.1–C9.6.3;
SUPPLEMENT C9.6.4

1

a Which of the following metals is found in nature as a pure element? Circle the correct letter.

- **A** magnesium
- **B** zinc
- **C** calcium
- **D** silver [1]

b Give a reason for your answer.

.... [1]

2 Which of the following statements about the extraction of iron is **untrue**? Circle the correct letter.

- **A** Iron ore is called hematite.
- **B** Iron ore is also known as iron(II) oxide.
- **C** The extraction of iron is a redox process.
- **D** Carbon is a key element in the extraction of iron. [1]

3

a) Aluminium is extracted from a mineral containing aluminium. What is the name of this mineral?

.... [1]

b) State the name of the process in which aluminium is extracted from its mineral.

.... [1]

c) Explain why aluminium cannot be extracted from its ore using carbon.

.... [1]

4 In a blast furnace:

SUPPLEMENT

a) Name the **three** solid substances that are added to the furnace.

1 ....

2 ....

3 .... [3]

b) **i** Air is blasted into the furnace and reacts to form a gas. What is this gas?

.... [1]

ii This gas is then reduced to form another gas. What is this gas?

.... [1]

iii Write a word equation for the reaction of the gas you have named in part **ii** with iron(III) oxide.

.... [1]

iv In the word equation in part **iii**, what is the name of the process the iron(III) oxide undergoes?

.... [1]

5 Write chemical equations for the following reactions involved in the extraction of iron in the blast furnace:
SUPPLEMENT

a The formation of carbon monoxide

.... [2]

b The reaction of carbon monoxide with iron(III) oxide

.... [2]

C10 Chemistry of the environment
Water

Student's Book pages 390–392 | Syllabus learning objectives C10.1.1–C10.1.3

1 The domestic water we use in our homes is treated in a number of stages.

   a What process is used to remove solids?

.... [1]

   b What is added to remove odours?

.... [1]

   c Chlorine is also used. What does the chlorine do?

.... [1]

2 Anhydrous copper(II) sulfate can be used to detect the presence of water.

   a What does the word anhydrous mean?

.... [1]

   b If water is present, what colour change will occur when adding the liquid to the anhydrous copper(II) sulfate?

....

.... [2]

3 A chemist made a sodium chloride solution by dissolving solid sodium chloride in water. Explain why they used distilled water rather than tap water.

....

.... [2]

Air quality and climate

Student's Book pages 392–398 | Syllabus learning objectives C10.2.1–C10.2.3; **SUPPLEMENT** C10.2.4–C10.2.6

1 What is the percentage of oxygen in the air? Circle the correct letter.

  A 21%

  B 15%

  C 78%

  D 10% [1]

> **TIP**
> Most of the gases that are common pollutants are oxides of non-metals.

2 In which of the following situations will methane be released into the atmosphere? Circle the correct letter.

  A burning petrol in a car engine

  B the incomplete burning of natural gas

  C from the digestive systems of animals like cows

  D lightning strikes in the atmosphere [1]

3 When carbon in the form of coal is burnt:

  a Which gas is formed if there is plenty of oxygen?

  .... [1]

b Which gas is formed if the amount of oxygen is very small?

.... [1]

4 Most particulates originate from the exhaust fumes of cars or lorries.

a What are particulates?

.... [1]

b Name **two** problems that particulates can cause for people living near busy roads.

1 ....

2 .... [2]

5 Which of the following gases cause acid rain in the atmosphere? Circle the correct letter.

A carbon monoxide

B nitrogen

C sulfur dioxide

D neon [1]

6 Planting trees is one way of reducing the level of greenhouse gases in the atmosphere. Explain why planting trees has this effect.

SUPPLEMENT

....

.... [2]

7 Name **two** other ways to reduce the levels of greenhouse gases in the atmosphere.

1 .... [1]

2 .... [1]

8 **a** Describe how greenhouse gases cause global warming.

....

.... [2]

b Name a common greenhouse gas.

.... [1]

C11 Organic chemistry
Fuels

Student's Book pages 410–413 | Syllabus learning objectives C11.2.1–C11.2.6; **SUPPLEMENT** C11.2.7

1

a What name is given to a compound containing carbon and hydrogen only?

.... [1]

> **TIP**
> You may need to check the section on Atoms, elements and compounds in the Student's Book.

b **i** Will a compound containing carbon and hydrogen only be ionically or covalently bonded?

.... [1]

ii Explain your answer to part **i**.

.... [1]

2 The following diagram shows how crude oil is separated in industry.

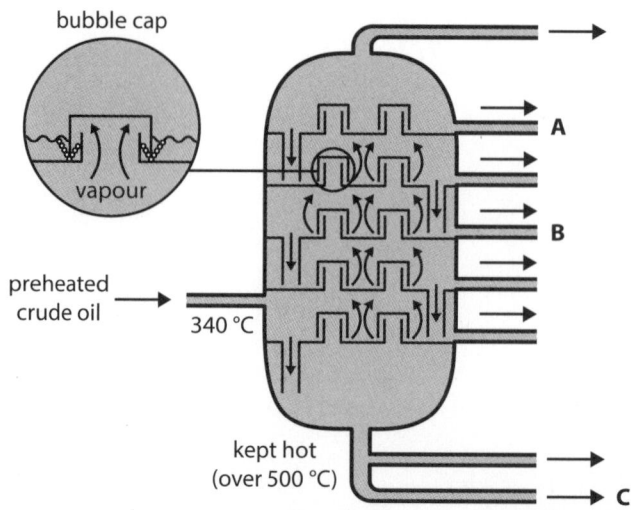

a State the name of the process for separating crude oil into its useful components.

.... [1]

b Three of the components produced in this process are petrol, diesel oil and bitumen. Match the names of these three components to the letters A, B and C in the diagram.

A ....

B ....

C .... [3]

c Complete the following table by giving a use for each of the three components.

Component	Use
Petrol	
Diesel oil	
Bitumen	

[3]

3 Methane (CH_4) is the main component of natural gas.

> **TIP**
> Remember what you learnt in the Chemistry of the environment topic!

a What problem is caused by high levels of methane in the atmosphere?

.... [1]

b i Write a word equation for the combustion of methane.

.... [1]

ii Write a balanced symbol equation for the combustion of methane.

.... [2]

C11 Organic chemistry | Alkanes

4 The following table shows some of the properties of the components formed in the separation of crude oil. Complete the table by placing ticks (✓) in the boxes to show which component has the higher value for that property.

Component	Chain length/number of carbon atoms per molecule	Boiling point
Petrol		
Bitumen		

[2]

Alkanes

Student's Book pages 416–419 | Syllabus learning objectives C11.1.1–C11.1.2; C11.3.1–C11.3.2; SUPPLEMENT C11.1.3–C11.1.4

1 Butane is a saturated hydrocarbon and an alkane.

a What is a hydrocarbon?

.... [1]

b What does the term saturated mean?

.... [1]

c Butane reacts with oxygen in a combustion reaction.

Write a word equation for the combustion of butane in a plentiful supply of oxygen.

.... [1]

2 a What is a homologous series?

....

.... [1]

b List **two** general characteristic properties of a homologous series.

1 ....

2 .... [2]

Alkenes

Student's Book pages 422–427 | Syllabus learning objectives C11.1.2; C11.4.1–C11.4.2; SUPPLEMENT C11.4.3–C11.4.4

1 a Ethene is an alkene. Is ethene a saturated or unsaturated hydrocarbon? Explain your answer.

....

.... [2]

b A test can be used to determine whether a hydrocarbon is saturated or unsaturated.

　i What chemical would you use in the test?

　.. [1]

　ii What observation would confirm the presence of an alkene?

　.. [1]

C11 Organic chemistry | Alkenes

2 The fractional distillation of crude oil produces a very low proportion of alkenes.

> **TIP**
> Fractional distillation was covered in the topic on alkanes.

a What is the name of the process that follows fractional distillation and produces alkenes from alkanes?

.... [1]

b State **two** reaction conditions needed in this process:

1 ....

2 .... [2]

3 Alkenes undergo addition reactions.

a What is an addition reaction?

.... [1]

b Ethene will undergo an addition reaction with hydrogen.

i What conditions are needed for this reaction?

.... [1]

ii Write a word equation for this reaction.

.... [1]

Polymers

Student's Book pages 430–432 | Syllabus learning objectives C11.5.1–C11.5.2

1 Define the term monomer.

.... [1]

2 Ethene molecules can be reacted together under suitable conditions to produce longer hydrocarbon molecules.

This is shown in the diagram below:

a Is ethene an alkane or an alkene?

.... [1]

b What is the name of the product of the reaction described above?

.... [1]

c Name the type of reaction that takes place.

.... [2]

C12 Experimental techniques and chemical analysis
Experimental design

Student's Book pages 440–441; 444 | Syllabus learning objectives C12.1.1–C12.1.2

1 The following diagram involves the filtration of a mixture.

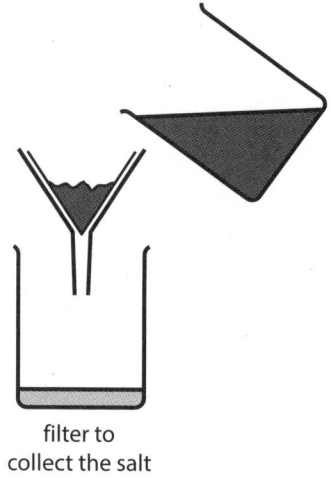

filter to collect the salt

On the diagram, label the:

a residue [1]

b filtrate [1]

2 During the preparation of a salt, a solution of the salt is evaporated until crystals form when a few drops of the solution are cooled.

a State how a solution is formed.

[2]

b What name is given to a hot solution that forms crystals when cooled?

.... [1]

> **TIP**
> You will be familiar with the next experiment if you have completed the Rate of reaction topic.

3 The following diagram shows how the rate of reaction between solid calcium carbonate solid and hydrochloric acid varies with the concentration of the hydrochloric acid used.

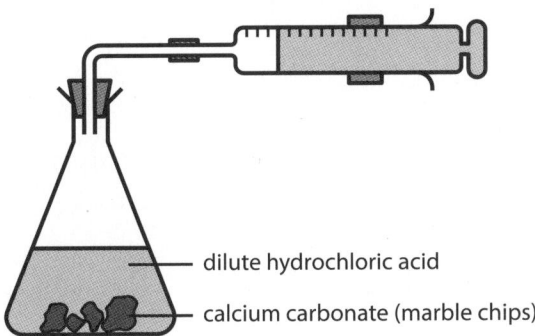

Four experiments are performed using the quantities shown in the following table.

Experiment	Mass of calcium carbonate (g)	Volume of 2 M hydrochloric acid (cm³)	Volume of distilled water (cm³)	Volume of gas produced in 30 seconds (cm³)
1	2	20	30	30
2	2	25	25	36
3	2	30	20	44
4	2	40	10	60

a What apparatus is used to collect the gas?

.... [1]

b What would you use to measure the mass of calcium carbonate used?

.... [1]

c What apparatus would you use to measure the time?

.... [1]

d **i** You have the choice of using a 100 cm³ measuring cylinder or a burette to measure the volumes of hydrochloric acid and distilled water.

Which would you choose to measure the volume accurately?

.... [1]

ii Explain your answer.

.... [1]

e To ensure a fair test, a student decides to measure the temperature of the hydrochloric acid solution before it is used. What apparatus should be used?

.... [1]

Chromatography

Student's Book pages 442–443 | Syllabus learning objectives C12.2.1–C12.2.2; SUPPLEMENT C12.2.3

1 Paper chromatography has been used to find out how many dyes there are in a sample of black ink. The apparatus and the chromatogram formed in the experiment are also shown below.

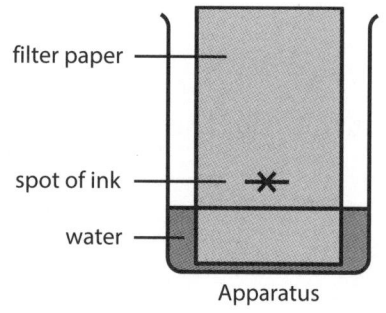
Apparatus

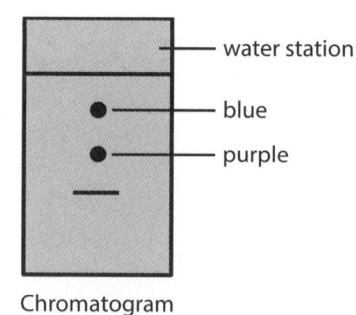
Chromatogram

a Why is the water level in the beaker below the spot of ink?

.... [1]

b No ink is left on the cross after 20 minutes when the water has soaked up to almost the top of the filter paper. Explain what this tells you about the black ink.

.... [1]

c Is the black ink a mixture or a pure substance? Explain your answer.

....

.... [2]

2 Paper chromatography is used to identify the components of a dye labelled M on the following chromatogram.

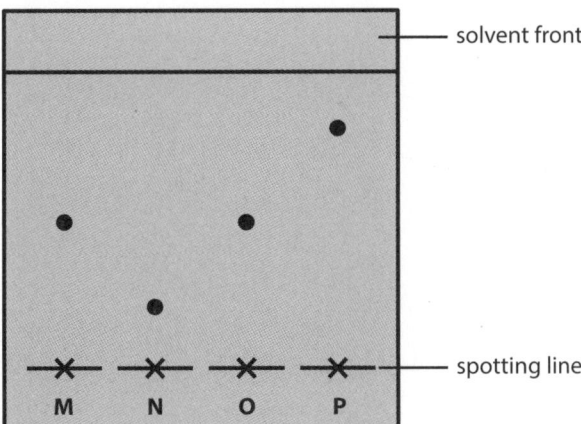

Which of the following statements is correct? Circle the correct letter.

    A The dye M contains substances N, O and P.

    B The dye M contains substances O and P.

    C The dye M contains substance O.

    D The dye O is the most soluble substance in the solvent used. [1]

3 a The R_f value of a component can be used to identify a substance. State the formula used to calculate an R_f value.

.... [1]

b Study the following chromatogram.

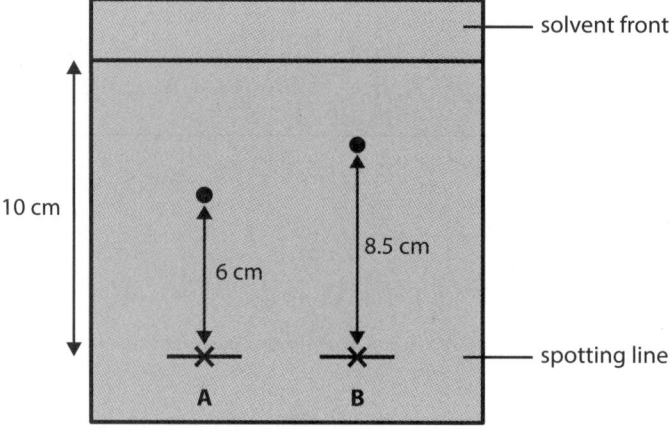

i What is the R_f value of substance A?

....

.... [1]

ii What is the R_f value of substance B?

....

.... [1]

Separation and purification

Student's Book pages 444–446 | Syllabus learning objectives C12.3.1–C12.3.2

1 For each of the following mixtures (**a**–**d**), state the method of separation that could be used.

a Salt solution and sand

.... [1]

b Common salt and lumps of marble (calcium carbonate)

.... [1]

c Sodium chloride from a saturated solution of sodium chloride

.... [1]

d Water from a solution of sodium chloride

.... [1]

2 A hot solution of sodium chloride solution is left to crystallise. What processes are needed for crystals of sodium chloride to form?

....

.... [2]

3 Why is fractional distillation used in some separations rather than simple distillation?

....

.... [1]

4 Look at the following apparatus.

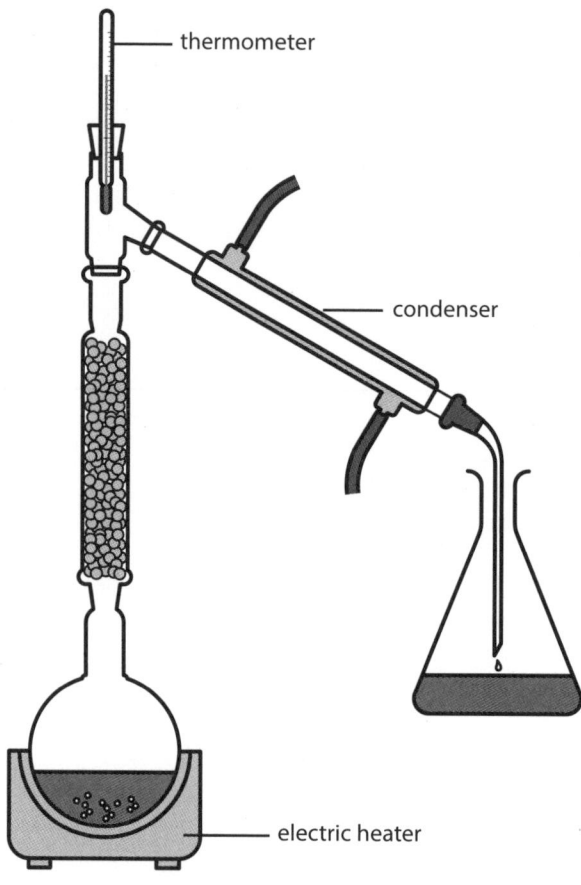

a What is the name of the process this apparatus is used for?

.... [1]

b Name **two** substances that could be separated using this apparatus:

1 ....

2 .... [2]

c Explain why this method separates these two substances.

.... [1]

d Which industrial process uses this method?

.... [1]

> **TIP**
> You will have covered this process in your study of fuels.

Identification of ions and gases

Student's Book pages 449–455 | Syllabus learning objectives C12.4.1–C12.4.4

1 Which of the following describes solutions contains Br⁻ ions? Circle the correct letter

   A It forms a white precipitate with dilute nitric acid and silver nitrate.

   B It forms a white precipitate with dilute nitric acid and barium nitrate.

   C Red litmus turns blue with warm sodium hydroxide solution and aluminium.

   D It forms a cream precipitate with dilute nitric acid and silver nitrate. [1]

2 Which of the following is a test for hydrogen gas?

   A It relights a glowing splint.

   B It produces a popping sound with a lighted splint.

   C It turns damp litmus paper blue.

   D It turns limewater cloudy.

.... [1]

C12 Experimental techniques and chemical analysis | Identification of ions and gases

3

a Describe how to perform a flame test.

[3]

b What colour flame would be produced on testing sodium chloride?

[1]

4 Some dilute sodium hydroxide is added to a solution of copper(II) sulfate.

a What would you expect to observe?

[1]

b What would you observe if dilute ammonia solution was added to copper(II) sulfate followed by excess dilute ammonia solution?

[2]

> **TIP**
>
> A cation is a positive ion which would be attracted to the cathode in electrolysis.

5 Complete the following table which shows the results of adding a test chemical/reagent to a solution containing a cation.

Cation	Test chemical	Result
Fe^{2+}		Green precipitate forms
	Sodium hydroxide solution added	Reddish-brown precipitate forms
Zn^{2+}		White precipitate forms but dissolves when excess chemical is added
NH_4^+		Damp red litmus turns blue

[4]

6 You are provided with a white powder sample which you think may be ammonium carbonate. Complete the following table with the results that would confirm that the powder *is* ammonium carbonate.

Ion	Test	Result
NH_4^+		
CO_3^{2-}		

[4]

Physics

P1 Motion, forces and energy — **134**
Physical quantities and measurement techniques — 134
Speed — 136
Distance–time graphs — 138
Acceleration — 140
Mass and weight — 143
Density — 144
Determining density — 147
Forces — 149
SUPPLEMENT Understanding the equation $F = ma$ — 151
Energy — 153
Energy resources — 155
Work done and power — 157
Pressure — 160

P2 Thermal physics — **163**
Kinetic particle model of matter — 163
Thermal expansion — 165
Evaporation — 168
Conduction — 169
Convection — 171
Radiation — 172
Consequences of thermal energy transfer — 175

P3 Waves — **177**
General properties of waves — 177
Reflection of light — 180
Refraction of light — 182
Lenses and dispersion — 183
Electromagnetic spectrum — 185
Sound and ultrasound — 188

P4 Electricity — 192

Electrical charge	192
Electric current	193
Electromotive force and potential difference	195
Resistance	197
Energy and power	199
Circuit diagrams and components	201
Series and parallel circuits	203
Electrical safety	206
Cells, batteries, generators and motors	208

P5 Space physics — 209

The Solar System	209
Stars	210
Galaxies and the Universe	213

P1 Motion, forces and energy
Physical quantities and measurement techniques

Student's Book pages 466–468 | Syllabus learning objectives P1.1.1–P1.1.3

1 For each question, circle the correct answer.

a How many cm are there in 1 m? [1]

　　A 0.01　　B 0.1　　C 100　　D 10 000

b How many cm^3 are there in 1 m^3? [1]

　　A 10　　B 100　　C 1000　　D 1 000 000

2 A student measures the length of a magnet using a plastic ruler. The ruler is placed slightly away from the edge of the magnet and the student measures the length of the magnet from the position shown in the diagram.

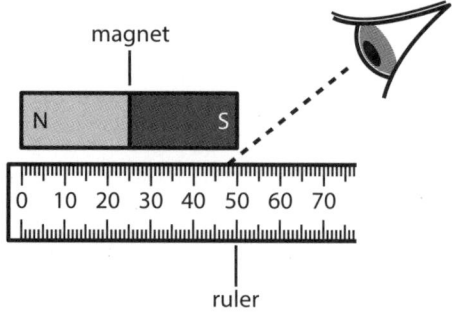

The student records the length of the magnet as 48 mm.

Explain why this measurement is incorrect and suggest the correct length of the magnet.

....

.... [2]

3 The diagram shows some water in a measuring cylinder.

a What is the volume of the water?
Give a reason for your answer.

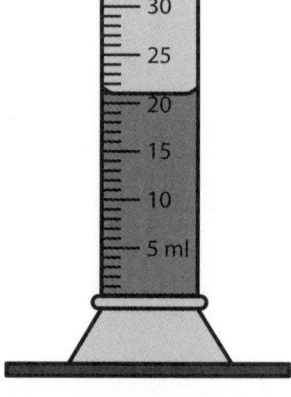

.... [2]

b A small solid metal cube is gently dropped into the water. It sinks to the bottom of the measuring cylinder. The new volume reading taken from the measuring cylinder is 29 ml, where 1 ml = 1 cm³.

Determine the volume of the metal cube in cm³.

volume = ... cm³ [1]

c Each side of the metal cube has length 2.0 cm.

  i Calculate the volume of the metal cube.

volume = ... cm³ [2]

  ii State how your value in **(c)(i)** compares with the value in **(b)**.

.... [1]

4 a In an experiment, a small metal ball is dropped from a great height into a tray of soft sand. The diameter D of the crater formed by the impact of the ball with the sand is measured to the nearest mm using a ruler.

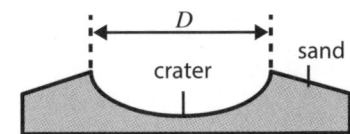

The experiment is repeated five times. The results are shown below.

Diameter D (mm)	12	10	19	11	12

  i Identify the most likely incorrect reading of the diameter.

.... [1]

  ii Determine the average diameter of the crater without the value identified in your answer above. Write your answer to the nearest mm.

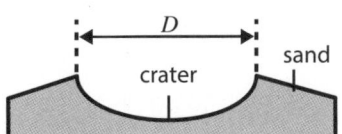

average diameter .... mm [2]

135

b The table below shows some results collected by a student investigating how the length *L* of a pendulum affects the time it takes to make 20 oscillations (or swings).

Length *L* (cm)	Time for 20 oscillations (s)	Period (s)
30.0	22.0	
60.0	31.1	

A stopwatch is used to time the 20 oscillations. The time for one complete oscillation is called the *period*.

i Complete the last column in the table. Write your answer to one decimal place. [2]

ii Use the table to suggest how the length affects the period.

.... [1]

iii Explain why it is sensible to determine the period by timing 20 oscillations rather than the timing a single oscillation.

....

.... [2]

Speed

Student's Book pages 471–473 | Syllabus learning objectives P1.2.1–P1.2.2

1 a Complete each equation by giving the name of the missing quantity.

i speed = distance ÷ .... [1]

ii distance = ... × time [1]

b A student wants to determine the average speed of a trolley rolling down a ramp from point A to point B.

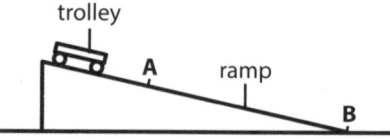

The student uses a stopwatch to determine the average time taken for the trolley to travel from A to B. The average time is 2.4 s. The distance between A and B is 0.60 m.

i How do you think that the student would have measured the distance between A and B?

.... [1]

ii Calculate the average speed of the trolley.

average speed = ... m/s [2]

2 a The table below shows the time taken by three cars A, B and C to travel a distance of 2.0 km.

Car	A	B	C
Time (minutes)	3.0	3.4	2.7

i Which car is the fastest? Explain your answer.

....

.... [2]

ii Use the table to determine the average speed of car A in m/s.

> **TIP**
> For the speed to be in m/s you need to first convert the distance from km to m. The kilo (k) prefix is equal to 1000 or 10^3. The time also needs to be converted in seconds. There are 60 s in one minute.

average speed = ... m/s [3]

b The speed of a motorboat is 34 m/s.

  i Calculate the distance travelled by the motorboat in a time of 20 s.

  distance = ... m [2]

  ii Calculate the time taken by motorboat to travel 10 m. Write your answer to two significant figures.

  time = ... s [2]

Distance–time graphs

Student's Book pages 473–476 | Syllabus learning objectives P1.2.3, P1.2.5 and P1.2.7

1 Two distance–time graphs for an object are shown below. For each distance–time graph, describe the motion of the object and give a reason for your answer.

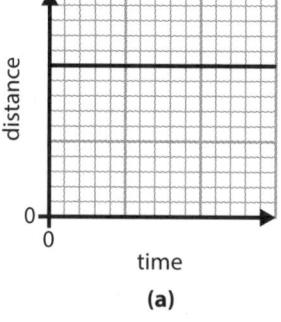
(a)

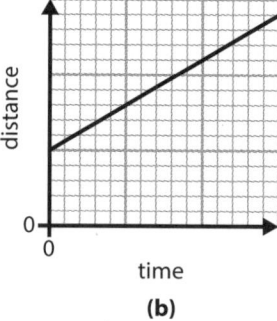
(b)

a ....

.... [2]

b ....

.... [2]

2 The distance–time graph for a walker is shown opposite.

a By examining the time intervals 0 s to 10 s and 10 s to 20 s, state when the walker is travelling the slowest.

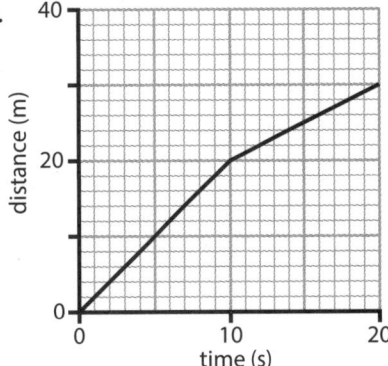

> **TIP**
> The command term 'state' requires a brief answer with no reasoning.

.... [1]

b Calculate the speed of the walker at time $t = 5.0$ s.

speed = ... m/s [2]

c How does the gradient of the line at time $t = 5.0$ s relate to your answer in **(b)**?

.... [1]

3 The distance–time graph for a car is shown opposite.

a Determine the total distance travelled by the car in 20 s.

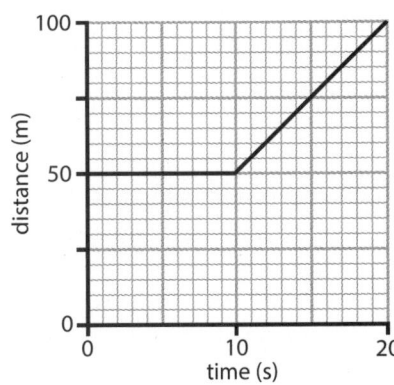

total distance = ... m [1]

b Determine the average speed v of the car over the 20 s interval.

$v =$ .... m/s [2]

c Calculate the speed of the car at 15 s.

speed = ... m/s [2]

Acceleration

Student's Book pages 475–479 | Syllabus learning objectives P1.2.3, P1.2.4 and P1.2.6; SUPPLEMENT P1.2.8–P1.2.13

1 An object falling on the surface of the Earth has a constant acceleration when the air resistance is negligible.

a SUPPLEMENT What is the approximate acceleration of free fall of an object falling near the surface of the Earth? Circle the correct letter.

    A 9.8 m/s   **B** 9.8 m/s^2 **C** 10 m/s   **D** 10 N   [1]

b i On the axes opposite, sketch the speed–time graph for an object released from rest and falling vertically downwards near the surface of the Earth.

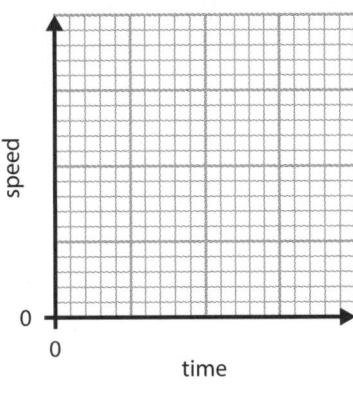

[2]

ii How would the speed–time graph above change when the object has an initial downward speed at the start? Explain your answer.

....

.... [2]

2 The speed–time graph for a cyclist is shown opposite.

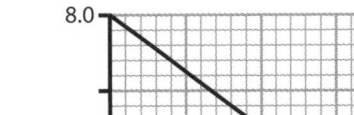

a Describe the motion of the cyclist.

....

.... [2]

b Calculate the total distance travelled by the cyclist from time $t = 0$ to time $t = 5.0$ s.

SUPPLEMENT

total distance = ... m   [4]

c **SUPPLEMENT** Use the graph to calculate the magnitude of the deceleration of the cyclist at time $t = 2.5$ s.

> **TIP**
> Deceleration is negative acceleration, which implies that the velocity decreases with time.

magnitude of deceleration = ... m/s² [3]

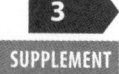 **SUPPLEMENT** A test rocket is fired upwards. It lifts up vertically and after a short period of time it falls freely under gravity back towards the surface of the Earth.

During one stage in its initial upward motion, the speed of the rocket changes from 100 m/s to 190 m/s in a time period of 3.6 s. Calculate the acceleration of the rocket.

acceleration = ... m/s² [3]

 **SUPPLEMENT** The magnitude of the constant deceleration of an object is 2.0 m/s². The initial speed of the object is 8.5 m/s.

Calculate its final speed after 3.0 s.

final speed = ... m/s [3]

Mass and weight

Student's Book pages 482–484 | Syllabus learning objectives P1.3.1–P1.3.3;
SUPPLEMENT P1.3.4–P1.3.5

1 a What is the unit of weight? Circle the correct letter.

    A metre   **B** second   **C** kilogram   **D** newton [1]

b Which statement below is correct for defining the mass of an object? Circle the correct letter.

    A Mass is a measure of how many atoms there are in an object.

    B Mass is a measure of the volume of an object.

    C Mass is a measure of the quantity of matter in an object.

    D Mass is a measure of the weight of an object. [1]

c Complete the sentence below.

SUPPLEMENT

The weight of an object is the effect of a .... field on

the .... [1]

d Circle two quantities below that are equivalent to each other.

SUPPLEMENT

mass   gravitational field strength   acceleration of free fall   weight [1]

2 a In words, define gravitational field strength g.

.... [1]

b Write an equation for gravitational field strength g in terms of gravitational force (or its weight) W acting on an object and its mass m.

.... [1]

P1 Motion, forces and energy | Density

c) The gravitational field strength close to the surface of the Earth is 9.8 N/kg. Complete the table opposite, where m is the mass of the object and W is the weight of the object at the Earth's surface. Write your values to two significant figures.

m / kg	F / N
45	
	1200
0.20	

[3]

3 Scientists used a space probe to determine the tiny force experienced by a small metal ball on a certain planet. The mass of the metal ball is 1.20 g and its weight is 0.030 N.

a) Calculate the gravitational field strength g.

$g = $ .... N/kg [3]

b) Suggest whether or not the space probe was on the surface of the Earth.

.... [1]

Density

Student's Book pages 486–487 | Syllabus learning objectives P1.4.1

1 a) Write an equation for density ρ of a material. Define any additional terms used.

....

.... [2]

b The density of water is 1.0 g/cm³.

What is the mass of the water with the following volumes?

i   volume = 2.0 cm³   mass = ... g   [1]

ii volume = 10 cm³   mass = ... g   [1]

iii volume = 1000 cm³   mass = ... g   [1]

c The density of a metal is 2.7 g/cm³.

Calculate the density of the metal in kg/m³.

> **TIP**
> There are 1000 g in 1 kg and there are 1 000 000 cm³ in 1 m³ or 1 cm³ = 10^{-6} m³

density = ... kg/m³   [3]

2 a Calculate the volume of a material given its density is 8.0 g/cm³ and it has mass of 96 g.

volume = ... cm³   [3]

b The density of a rock is 550 kg/m³ and it has a volume of 0.020 m³. Calculate the mass of the rock in kg.

mass = ... kg   [3]

c A column of liquid has cross-sectional area 3.2 cm² and height 15 cm. The mass of the liquid column is 36 g. Calculate the density of the liquid in g/cm³.

density = ... g/cm³   [3]

145

3 The table below shows the density of three metals.

A student has a rectangular block of metal. The length of the block is 4.0 cm, its width is 2.0 cm and its height is 2.5 cm. The mass of the block is 156 g.

Metal	Aluminium	Steel	Gold
Density (g/cm^3)	2.7	7.8	19.3

a Calculate the volume of the metal block.

volume = ... cm^3 [1]

b Determine the density of the metal and use the table above to identify the metal.

density = ... g/cm^3

The metal is .... [3]

Determining density

Student's Book pages 488–491 | Syllabus learning objectives P1.4.2–P1.4.3

1 Describe how you can determine the density of a wooden block in the shape of a cube. In your description, include:

- the equipment you would use
- the measurements you need to take
- how the data collected will be used to determine the density.

[4]

2 The information opposite is collected by a student who is determining the density of vegetable oil.

Mass of empty measuring cylinder	120 g
Volume of oil in measuring cylinder	42 cm³
Total mass of measuring cylinder with the oil	156 g

Use the information provided to calculate the density of the oil.

density = ... g/cm³ [3]

3 A student is given a measuring cylinder and a supply of water. A digital balance is also available.

Describe how the student can determine the density of the metal bolt shown below. In your description, include:

- the measurements you need to take
- how the data collected will be used to determine the density.

[4]

4 The density of wax is about 0.9 g/cm^3, and the density of water is 1.0 g/cm^3.

Explain whether or not the wax will float or sink in water.

[2]

Forces

Student's Book pages 494–498 | Syllabus learning objectives P1.5.1.1–P1.5.1.6

1 Various forces act on an object. Tick ✓ all the effects that these forces can have on the object.

Forces can change the object's …	Place a tick ✓ here if correct
colour	
shape	
size	
direction of travel	
speed	
weight	

[1]

2 A heavy box is pushed along a level ground at a constant speed. The force pushing the box to the right is 60 N.

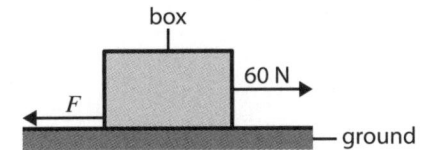

a Describe what is meant by solid friction in relation to the box.

.... [1]

b State the value of the frictional force F acting on the box.

$F = $ .... N [1]

c What is the resultant force acting on the box? Give a reason for your answer.

TIP
A non-zero resultant force is needed to change the speed of an object and/or its direction.

....

.... [2]

149

3 In the two examples below, each object is **falling** vertically downwards at a *constant* speed.

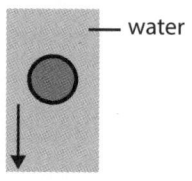

a Parachutist falling through air

b Ball falling in water

> **TIP**
> Drag refers to the frictional force acting on an object moving through a liquid or a gas.

In each case, show the directions of the weight and the drag using arrows. [2]

4 In each case below determine the magnitude, and direction if any, of the resultant force and state whether or not its speed will change.

a ....

.... [2]

b ....

.... [2]

c ....

.... [2]

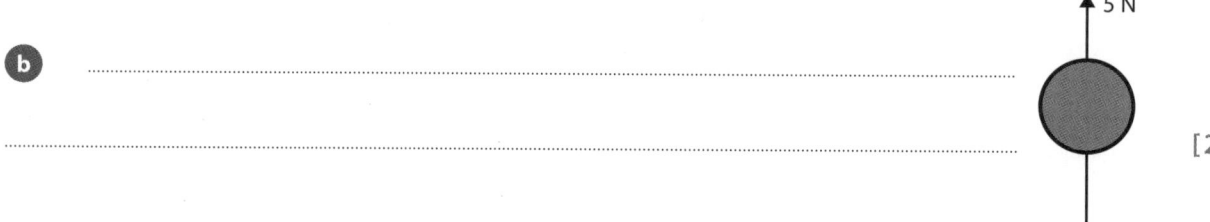

SUPPLEMENT: Understanding the equation $F = ma$

Student's Book pages 498–499 | Syllabus learning objective
SUPPLEMENT P1.5.1.7

1 A very useful equation in physics is $F = ma$.

a Identify all the terms in this equation.

F: .... m: .... a: .... [1]

b Complete the following two equations showing the relationship between the quantities identified in **(a)**.

$a =$ .... $m =$ .... [1]

c Which of the following is the correct alternative unit for the newton (N)? Circle the correct letter.

    A kg   **B** m/s²   **C** kg m s²   **D** kg m/s²   [1]

d The resultant force acting on a car is doubled. What happens to its acceleration? Give a reason for your answer.

....

.... [2]

e The direction of acceleration of an object is to the right. What is the direction of the resultant force?

.... [1]

2 The diagram below shows the horizontal forces acting on a 900 kg mass car when it is about to move off **(a)**, and then a short time later **(b)**.

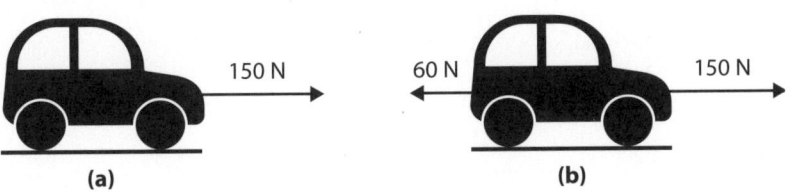

(a)   (b)

Calculate the acceleration of the car in each case.

a

acceleration = ... m/s² [2]

b

> **TIP**
> In the equation $F = ma$, F is the **total** or **resultant** force

acceleration = ... m/s² [3]

3 SUPPLEMENT The diagram opposite shows an air-powered toy rocket. The rocket lifts off vertically when the someone pushes on the foot pump.

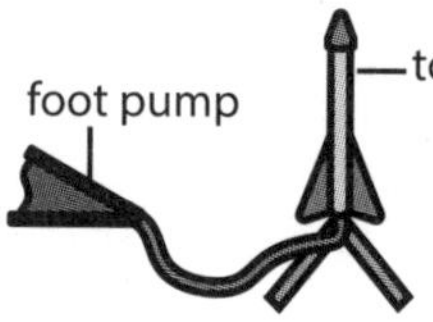

The mass of the rocket is 0.050 kg. The vertical upwards force on the rocket during lift-off is 0.70 N.

a Show that the resultant force on the rocket at lift-off is 0.21 N.

[2]

b Calculate the acceleration of the rocket at lift-off.

acceleration = ... m/s² [2]

Energy

Student's Book pages 502–506 | Syllabus learning objectives P1.6.1.1–P1.6.1.3; **SUPPLEMENT** P1.6.1.4–P1.6.1.5

1 Energy can be stored – the energy stores have different names. For example, when a rubber band is extended the energy stored within the rubber band is called elastic potential energy.

Complete the table below, by identifying the energy store for each description.

Description	Name of energy store(s)
Energy in cooking oil	
Energy of hot water to make tea or coffee	
Energy in the food we eat	
Energy in the nuclei of radioactive atoms	
Energy of a space probe moving between two planets	
Energy stored between an electron and a nucleus	

[7]

2 A high-speed train suddenly brakes and comes to a halt after travelling a short distance along the track. The wheels of the train skid along the track. Noise is generated as the train slows down. The wheels of the train and the track get very hot.

a Mechanical work is done by friction to slow down the train.

What is also produced in this process as energy is transferred between the energy stores identified in **(a)**? Circle the correct letter.

    A electrical   **B** light   **C** nuclear   **D** sound   [1]

b The simple energy flow diagram shows the energy transfers that take place when the train stops.

SUPPLEMENT

Insert the **two** missing energy stores to complete it.

Initial energy store		Final energy store
.... energy	Mechanical work done on the train	.... energy

[2]

3 A block of wood of 0.80 kg is placed on a ramp and released from rest from point A. It slides down the ramp picking up speed. At the bottom of the ramp, point B, the block has speed 2.0 m/s.

SUPPLEMENT

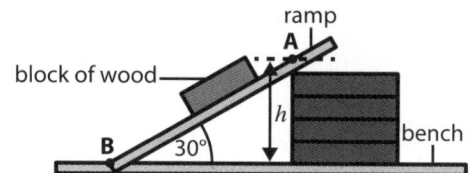

The distance between A and B is 1.5 m. The ramp makes an angle of 30 degrees to the horizontal.

a Show that the initial height h of the block is 0.75 m.

[2]

b Calculate the change in gravitational potential energy ΔE_p of the block as it slides from A to B.

$\Delta E_p = $ .... J   [3]

c Calculate the kinetic energy E_k of the block at the bottom of the ramp.

$E_k = $ .... J   [3]

154

d) Use the principle of conservation of energy to explain why the two values in **(b)** and **(c)** are not the same.

....

.... [2]

Energy resources

Student's Book pages 507–511 | Syllabus learning objectives P1.6.3.1;
SUPPLEMENT P1.6.3.3–P1.6.5.5

1 a) Biofuels, fossil fuels and nuclear fuels are used in the production of electricity in power stations. In these power stations in the first stage, water is heated in a boiler using energy from one of these fuels.

Complete the diagram below by identifying the last two stages in the production of electricity from the power station.

| boiler | → | .... | → | .... |

[2]

b) Complete the sentences below.

.... energy resources, when used, will be naturally replenished over time. An example of this type of energy resource is .... [2]

c) A non-renewable energy resource will be gone forever once used.

Tick ✓ all the non-renewable energy resources in the list below.

- biofuels
- fossil fuels
- electromagnetic waves from the Sun
- geothermal resources
- nuclear fuels

[1]

2 The diagram opposite shows a tidal power station at the mouth of a river meeting the sea (estuary). The turbine blades rotate and the generator produces electrical power when the tide comes in or when the tide goes out. The power station helps with the transmission of electrical power across the land.

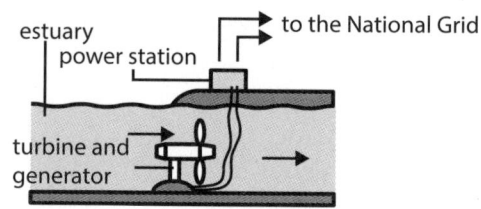

What is the input energy store for this tidal system? Circle the correct letter.

A kinetic

B internal (thermal)

C chemical

D nuclear

[1]

3 Complete the last column of the table below.

Energy resource	Energy store
Biofuel	
Fossil fuel (coal, gas and oil)	
Geothermal	
Hydroelectricity	
Nuclear fuel	
Solar	
Tidal	
Water waves	
Wind	

[9]

4 a Complete the sentence below:

SUPPLEMENT

The radiation from the Sun is the main source of energy for all our energy resources except for g.................................., n..................................
and t.................................. .

[3]

b) The race is on to produce large scale electricity using nuclear fusion. Very high temperatures are required to trigger nuclear fusion between hydrogen nuclei to produce helium nuclei. Some scientists think that this may happen by about 2040.

Name the hot glowing ball of gas that releases energy naturally from nuclear fusion.

.... [1]

Work done and power

Student's Book pages 511–515 | Syllabus learning objectives P1.6.2.1–P1.6.2.2; P1.6.3.2; P1.6.4.1; **SUPPLEMENT** P1.6.3.6

1 A vase is lifted from the floor onto a shelf. Which two quantities are required to calculate the work done on the vase? Circle the correct letter.

  A the mass of the vase and its volume

  B the mass of the vase and the time taken to lift the vase

  C the weight of the vase and the height of the shelf from the floor

  D the weight of the vase and the time taken to lift the vase [1]

2 a) Complete the sentence below.

The mechanical or electrical work done on an object is equal to the .... transferred. [1]

b) The unit of work done is the joule (J). Which of the following is correct? Circle the correct letter.

  A 1 J = 1 N m

  B 1 J = 1 kg m

  C 1 J = 1 N/m

  D 1 J = 1 kg m/s^2 [1]

P1 Motion, forces and energy | Work done and power

c The force acting on an object is F and the distance moved by the object in the direction of the force is d. Complete the missing items in the table below.

F (N)	d (m)	Work done (J)	Energy transferred (J)
100	50		
	160	8000	
45		270	

[6]

3 **a** Define power in terms of work done.

.... [1]

b The unit of power is the watt (W). Which of the following is correct? Circle the correct letter.

A 1 W = 1 J s

B 1 W = 1 J/s

C 1 W = 1 N s

D 1 W = 1 N/s

[1]

c A car is travelling on a straight road. The driver applies the brakes. The kinetic energy of the car changes from 200 kJ to 50 kJ in 7.5 s.

i Calculate the braking power of the car.

power = ... W [3]

ii What energy store is the kinetic energy of the car transferred to?

.... [1]

4 The output power of a water pump is 200 W.

a What is the energy transferred per second by the water pump?

.... [1]

b Calculate the energy transferred by the water pump in a time of 30 minutes.

> **TIP**
> You must convert the time into seconds. There are 60 s in one minute.

energy transferred = ... J [2]

5 The falling water from a dam is used to generate electrical power.

In a time of 1.0 minutes (60 s), the mass of water falling through a height of 20 m onto the turbine blades of the generator at the bottom of the fall is 30 000 kg.

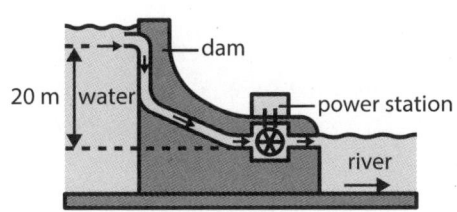

a Show that the weight of 30 000 kg of water is 294 000 N. [1]

b Calculate the work done by the force of gravity on the 30 000 kg mass of water falling through 20 m.

work done = ... J [2]

c Calculate the total power input to the generator.

total power input = ... W [2]

d Explain what is meant by the statement 'the efficiency of the power station is not 100%'.

[1]

6 a (SUPPLEMENT) At present, some electricity is produced by nuclear fission power stations. The efficiency of a typical power station producing electricity is about 35%.

A nuclear power station produces a power output of 1.2 GW. Calculate the total power input. Give your answer in GW, where 1 GW = 10^9 W.

total power input = ... GW [3]

b A geothermal power station produces 30 000 J in a particular time. The total energy input to the power station in the same period of time is 375 000 J. Calculate the efficiency of this type of power station.

efficiency = ... % [3]

Pressure

Student's Book pages 520–522 | Syllabus learning objectives P1.7.1–P1.7.2

1 a Write an equation for pressure p in terms of the force F and the area A.

.... [1]

b In an experiment, a student has the force acting on a surface in newtons (N) and the area over which the force acts in cm².

The pressure is calculated as 1.5 N/cm². What is this pressure in N/m²?

> **TIP**
> There are 100 cm in 1 m, therefore there are 10^4 cm² in 1 m².

pressure = ... N/m² [2]

c All the objects below are drawn to scale, and each object has the **same** weight.

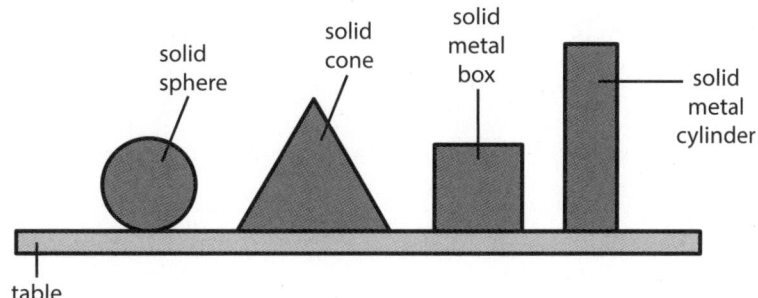

Explain which object exerts the least pressure on the table.

....

.... [2]

d Explain why is easier to cut bread, or fruit, with a sharp knife rather than a blunt knife.

....

.... [2]

2 The weight of a person is 520 N when they are standing on soft sand. The **total** surface area of the shoes the person is wearing is 140 cm^2.

a Calculate the pressure p exerted by the person on the soft sand.

$p =$ .... N/cm^2 [2]

b Without any further calculation, explain the effect on the pressure exerted when the person stands on just one leg.

....

.... [2]

3 A concrete column in the shape of a long cylinder is resting on the ground as shown. The cross-sectional area of the concrete column is 0.52 m², and its mass is 9.0 tonnes. (1 tonne = 1000 kg).

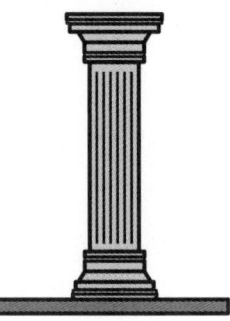

Calculate the pressure, in N/m², exerted by the concrete column on the ground.

pressure = ...N/m²   [3]

P2 Thermal physics
Kinetic particle model of matter

Student's Book pages 532–537 | Syllabus learning objectives P2.1.1.1–P2.1.1.2; P2.1.2.1–P2.1.2.2; SUPPLEMENT P2.1.2.3–P2.1.2.4; P2.1.3.1

1 The table below has some descriptions that relate to the three states of matter – solid, liquid and gas. Complete the last column by identifying the state of matter. The first one has been done for you.

Description	State of matter
The particles (atoms) are arranged in a regular pattern and do not move apart from each other.	solid
Particles move very fast with a range of speeds and directions.	
The substance will take the shape and volume of the container.	
A solid will change to this state of matter when its temperature is increased.	
The particles vibrate about fixed positions.	
A gas will change to this state of matter when its temperature is decreased.	
The substance can be compressed because there is lots of empty space between the particles.	

[6]

2 A lump of ice is placed in a pan and heated.

a Name the change in state that takes place as the ice turns into liquid water.

.... [1]

b State what happens to the motion of particles as this change happens.

.... [1]

P2 Thermal physics | Kinetic particle model of matter

c The water continues to be heated. Describe the changes to the motion and separation of the particles as the liquid water turns into a gas (steam).

....

.... [2]

3 A container has a fixed amount of gas. Using the words below, explain how the gas particles (atoms) exert pressure on the container walls.
SUPPLEMENT

        force   collide   area

....

....

....

.... [4]

4 A small metal container has a fixed amount of air. The temperature of the air inside is decreased.

a Suggest how you could decrease the temperature of the container in the laboratory.

.... [1]

b State whether the speed of the air particles decreases, increases or stays the same.

.... [1]

c State whether the pressure exerted by the air decreases, increases or stays the same.
SUPPLEMENT

.... [1]

5 **SUPPLEMENT** A small rubber ball has some trapped air inside. The volume of the ball is decreased by pushing the ball against a wall.

Which statement below best describes why the pressure of the air inside the ball increases? Circle the correct letter. [1]

    A   The air particles collide less frequently with each other.

    B   The air particles collide with the inside of the ball more frequently.

    C   The air particles move faster.

    D   The air particles move further apart.

6 **SUPPLEMENT** A helium-filled weather balloon is released from the surface of the Earth. The pressure exerted by the atmosphere on the balloon decreases as it climbs higher.

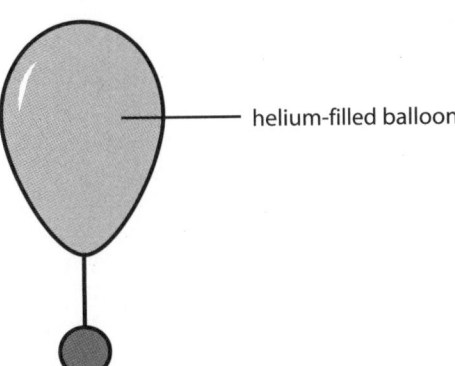

helium-filled balloon

Explain what happens to the volume of the balloon as it climbs higher, assuming the temperature inside the balloon remains constant.

....

.... [2]

Thermal expansion

Student's Book pages 540–542 | Syllabus learning objectives P2.2.1.1; **SUPPLEMENT** P2.2.1.2

1 **a** Circle the material that would expand the most, at constant pressure, for a given increase in temperature.

        steel   water   brick   air   ice   [1]

b) The table below has some statements related to thermal expansion. In the last column, state whether the statement is true or false.

Statement	True or false?
For a given increase in temperature, a liquid will expand much more than a gas.	
A metal bridge will be slightly longer during daytime than at night time.	
The volume of a gas, at a constant pressure, will increase when its temperature is increased.	
Increasing the temperature of a gas, at constant pressure, will increase the separation between the gas particles (atoms).	
All materials expand because the particles (atoms) that make them up become larger.	

[5]

2 At room temperature, a steel rod cannot be slotted into a steel washer. The hole in the washer is slightly smaller than diameter of the steel rod. However, when the washer is heated to a very high temperature, the washer easily slips over the steel rod (which is at room temperature). Explain why this is so.

....

....

[2]

3 SUPPLEMENT All materials expand when heated. A long brick wall often has a gap. This gap is filled up with a flexible material. Describe what may happen to the wall if the gap was not there.

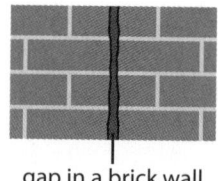

gap in a brick wall

....

....

[2]

4 a This question is about two different metal strips (steel and brass) of similar width and length, joined together. This arrangement is known as a bimetallic strip.

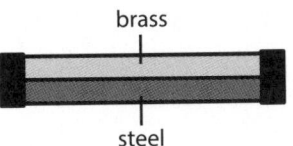

For a given change in temperature, the volume of brass increases almost twice as much as steel.

Draw a diagram of the bimetallic strip when its temperature is increased. Explain the shape of the strip.

....

....

.... [3]

b A copper pipe has length 5.20 m at 20 °C. Its length increases by 0.0017% for every 1 °C increase in temperature.

Calculate the change in the length ΔL of the pipe when it has hot water at 90 °C flowing through it. Give your answer in standard form.

$\Delta L = $ .... m [3]

Evaporation

Student's Book page 543 | Syllabus learning objectives P2.2.2.1–P2.2.2.2; **SUPPLEMENT** P2.2.2.3

1 Place a tick in the final column of the table if the statement is correctly describing evaporation.

Statement: During evaporation …	
cooling takes place.	
the more energetic particles escape the surface of the liquid.	
there is a change of state without change in temperature.	
of water, the temperature is always 100 °C.	

[2]

2 **SUPPLEMENT** **a** Boiling water is poured into a test tube. The water takes a very long time to cool down to room temperature.

State **two** things that you could do to increase the rate of cooling. Give a reason for each of your answers.

1 ....

....

2 ....

....

[4]

3 One way to cool your feet on a hot day is to spray some water onto your feet.

a Name the process that is responsible for this cooling effect.

.... [1]

b Explain how this process leads to the cooling of the skin.

....

....

.... [3]

Conduction

Student's Book pages 546–548 | Syllabus learning objectives P2.3.1.1; SUPPLEMENT P2.3.1.2

1 a (Circle) the **two** best thermal conductors from the list below.

air   copper   rubber   steel   wood   wool   [1]

b Suggest a common feature of the materials identified as the best thermal conductors.

.... [1]

c What term can you use as the opposite of good conductors? Give one example of such a material.

....

.... [2]

d Explain why it is sensible to make a cooking pot from a metal and its handle from either plastic or wood.

....

.... [2]

e In an experiment, two different test tubes are heated. Both test tubes have solid butter at the bottom. The test tube labelled A has water, and test tube B has metal pellets. The test tubes are heated from the top end as shown below.

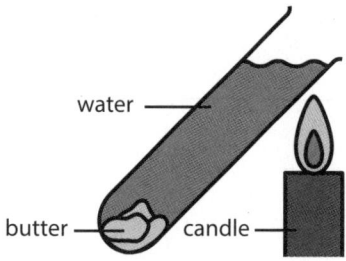 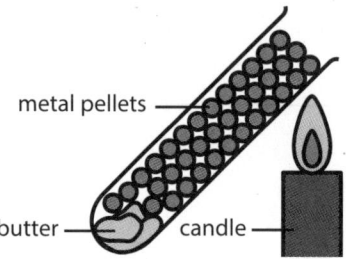

Describe in which test tube the butter would start to melt first.

....

.... [2]

2 a Which following term, or terms, fully explain the thermal conduction in metals? Circle the correct letter.

SUPPLEMENT

A lattice vibrations

B lattice vibrations, movement of delocalised (mobile) electrons

C lattice vibrations, movement of protons

D movement of delocalised (mobile) electrons [1]

b The diagram opposite shows particles within a metal. One end of the metal is hot and the other end is cold. Thermal energy is either transferred from A to B, or from B to A.

Identify the direction in which thermal energy is transferred.

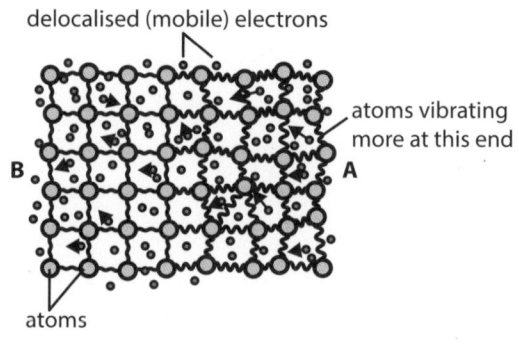

.... [1]

c) State why a metal such as copper is a better thermal conductor than a solid such as wood.

.... [1]

Convection

Student's Book pages 548–549 | Syllabus learning objectives P2.3.2.1–P2.3.2.2; **SUPPLEMENT** P2.3.2.3

1 Complete the sentence below.

Convection is an important method of thermal energy transfer in .... and .... In ...., convection cannot take place because the particles are held tightly together. [3]

2 This question is about convection in a liquid such as water.

SUPPLEMENT

Place a tick under the appropriate column for each quantity when water is heated in a beaker with a Bunsen burner.

What happens to …	Increases	Decreases	Stays the same
the average separation between water molecules?			
the density of water?			
the mass of the water?			
the temperature of the water?			

[4]

3 A light piece of paper is held over a table lamp. The bulb of the lamp is very hot. Explain why the paper over the lamp will be lifted upwards.

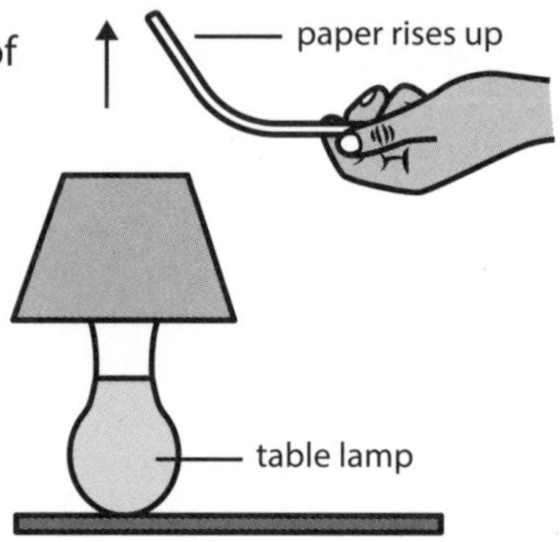

....

....

.... [2]

Radiation

Student's Book pages 549–552 | Syllabus learning objectives P2.3.3.1–P2.3.3.2;
SUPPLEMENT P2.3.3.3–P2.3.3.5

1 a The Sun is a hot glowing ball of gas in space. Almost all of the energy arriving at the Earth's surface comes from the Sun.

What is the method of transfer of energy from the Sun to the Earth? Circle the correct letter.

    A conduction   **B** convection   **C** nuclear   **D** radiation   [1]

b The diagram opposite shows a small camp fire.

> **TIP**
> Air is a poor conductor of thermal energy.

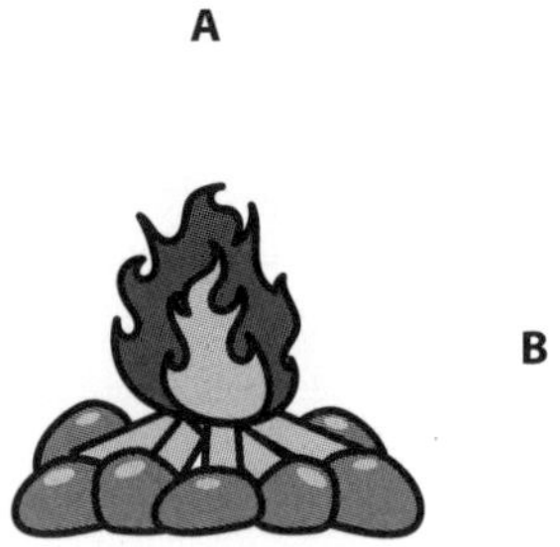

State the methods of heat transfer to points A and B.

....

.... [2]

c Thermal radiation is emitted by all objects. Circle the correct name of the waves responsible for this radiation.

    infrared   radio waves   sound   visible light   water waves   [1]

2 A student is investigating the absorption of infrared radiation by using three identical thermometers. The bulbs of the thermometers are painted in different colours – black, shiny white and grey.

All the thermometers are left in the room for some time and then hung out in bright sunshine for one hour. The temperatures from all three thermometers are recorded.

a Suggest why it was sensible to leave the thermometers in a room before taking them out into the bright sunshine.

.... [1]

b Explain the relative temperatures of the three thermometers.

....
....
....
.... [4]

3 One of the heating systems used in cold countries is a large liquid-filled metal tank. The tank has water, or oil, that is electrically heated during daytime. At night time, the tank is used to warm a room.

Explain the colour of the tank you would choose.

....
.... [2]

4 a A metal pen is placed under a hot table lamp. The pen starts to get warm and after some time reaches a constant temperature.

SUPPLEMENT

Explain why the temperature of the pen stays constant.

....
.... [2]

b The Earth receives infrared radiation from the Sun. The Earth also emits infrared radiation away from its surface. The graph opposite shows the variation of the temperature θ of the ground in a particular town after sunrise.

Explain why the temperature:

i rises after sunrise

....

.... [2]

ii decreases after sunset.

....

.... [2]

5 In an investigation on the emission of infrared radiation, a student is given two identical metal bottles – one is painted shiny white and the other is black.

Describe how you can demonstrate that the black bottle is a good **emitter** of infrared radiation. In your description, include:

- any additional equipment required
- the measurements you would take
- how the data collected will be used to reach a conclusion.

....

....

....

.... [4]

Consequences of thermal energy transfer

Student's Book pages 552–554 | Syllabus learning objectives P2.3.4.1

1 A metal cooking pan is placed over a fire. The pan contains water and some potatoes. The handle of the pan is made of metal.

a State the method by which

   i thermal energy is transferred through the bottom of the pan

   .... [1]

   ii the water gets hot

   .... [1]

   iii the potatoes get cooked.

   .... [1]

b Suggest why it is not sensible to touch the end of the pan handle with bare hands while the potatoes are cooking.

   ....

   .... [2]

P2 Thermal physics | Consequences of thermal energy transfer

2 The diagram shows a room with a heater at one end. The arrows show the direction of the movement of the air within the room when the window is closed.

a Explain the movement of the air in the room.

....

....

....

.... [4]

b It is much colder outside the room. Suggest why it would not be sensible to open the window.

....

.... [1]

3 A vacuum flask is used to keep hot liquid hot or cold liquids cold. The diagram shows the inner parts of a vacuum flask.

Explain how the liquid inside the flask is hot for longer because of the silver (aluminium) coating of the double glass walls and the vacuum between the glass walls.

....

....

.... [3]

176

P3 Waves
General properties of waves

Student's Book pages 562–567 | Syllabus learning objectives P3.1.1–P3.1.5;
SUPPLEMENT P3.1.6–P3.1.7

1 A long spring is stretched out on level-ground and fixed at one end. The other end of the spring is repeatedly shaken at right angles to the spring.

a State the type of wave motion this will generate.

.... [1]

b State what the vibrations transfer from one end of the spring to its other end.

.... [1]

c The diagram below shows the displacement–distance graph for the wave on the spring.

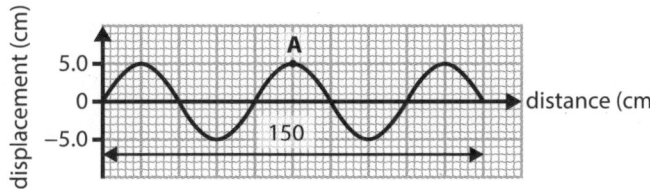

Determine the amplitude and wavelength of the wave.

> **TIP**
> The wavelength is the distance between adjacent peaks (or troughs).

amplitude = ... cm

wavelength = ... cm [3]

d The frequency of the wave is 3.0 Hz. What does this mean in terms of the motion of point A?

.... [1]

e *SUPPLEMENT* Describe how you can produce another type of wave motion on the stretched spring. In your description name the type of motion.

....

.... [2]

f *SUPPLEMENT* (Circle) all the waves that can be described as being transverse waves.

water waves   sound   seismic P–waves   light   seismic S–waves   [1]

2 Water waves in a ripple tank have wavelength 1.2 cm and frequency 20 Hz.

a Calculate the speed v of the water waves.

TIP
The speed is required in cm/s, so leave the wavelength in cm.

$v = $ .... cm/s   [3]

b The frequency of the waves is increased. Explain the effect this would have on the speed of the waves and their wavelength.

....

....

.... [3]

3 a Calculate the frequency f of a wave with speed 3000 m/s and wavelength 6.0 mm.

TIP
1 mm = 0.001 m or 10^{-3} m.

$f = $ .... Hz   [3]

b There are two types of wave motion – longitudinal waves and transverse waves. What is common for **both** of these types of waves? Circle the correct letter.

SUPPLEMENT

A They are vibrations in space.

B They transfer matter between two points.

C Vibrations are parallel to the direction of propagation.

D Vibrations are perpendicular to the direction of propagation. [1]

4 a The diagram shows a ripple tank used to demonstrate the properties of water waves.

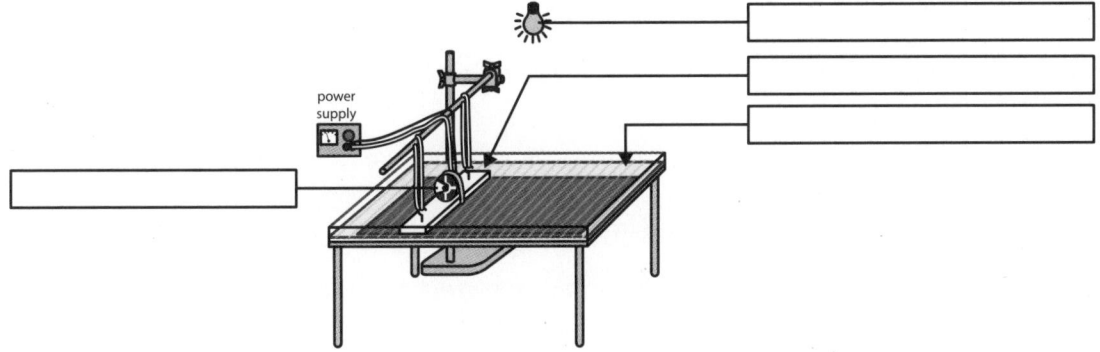

On the diagram, add the four missing labels for the ripple tank. [4]

b The following four diagrams show what happens to parallel wavefronts of water waves in a ripple tank. The waves are generated by a long-straight 'dipper' vibrating on the surface of the water.

Identify the effect being shown in each diagram and in each case state whether or not the speed of the wave changes.

i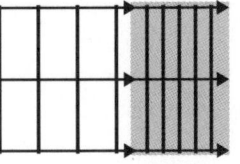

[2]

ii

[2]

Reflection of light

Student's Book pages 570–573 | Syllabus learning objectives P3.2.1.1–P3.2.1.3;
SUPPLEMENT P3.2.1.4–P3.2.1.5

1 The diagram opposite shows a ray of light from a laser that is incident at a plane mirror.

Complete the diagram by:

- drawing a normal
- drawing the reflected ray
- marking the angle of incidence i and the angle of reflection r.

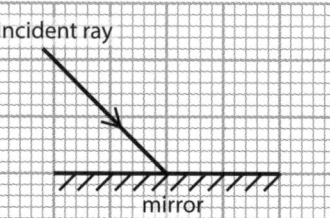

[3]

2 A person is looking at their reflected image in a plane mirror.

The following list shows some statements that may be correct about the characteristics of the image.

Tick ✓ the ones that are correct in this situation.

Possible characteristics of the image	Tick here if correct
The image is larger than the object.	
The image is same distance behind the mirror as the object is in front.	
The image is same size as the object.	
The image is laterally inverted (left appears as right).	
The image is on the surface of the mirror.	

[3]

3 An incomplete ray diagram is shown for locating the image of an object formed in a plane mirror.

SUPPLEMENT

a Complete the ray diagram. On the diagram indicate:

- the location of the image
- the height of the image
- where you would place your eye to see the image. [4]

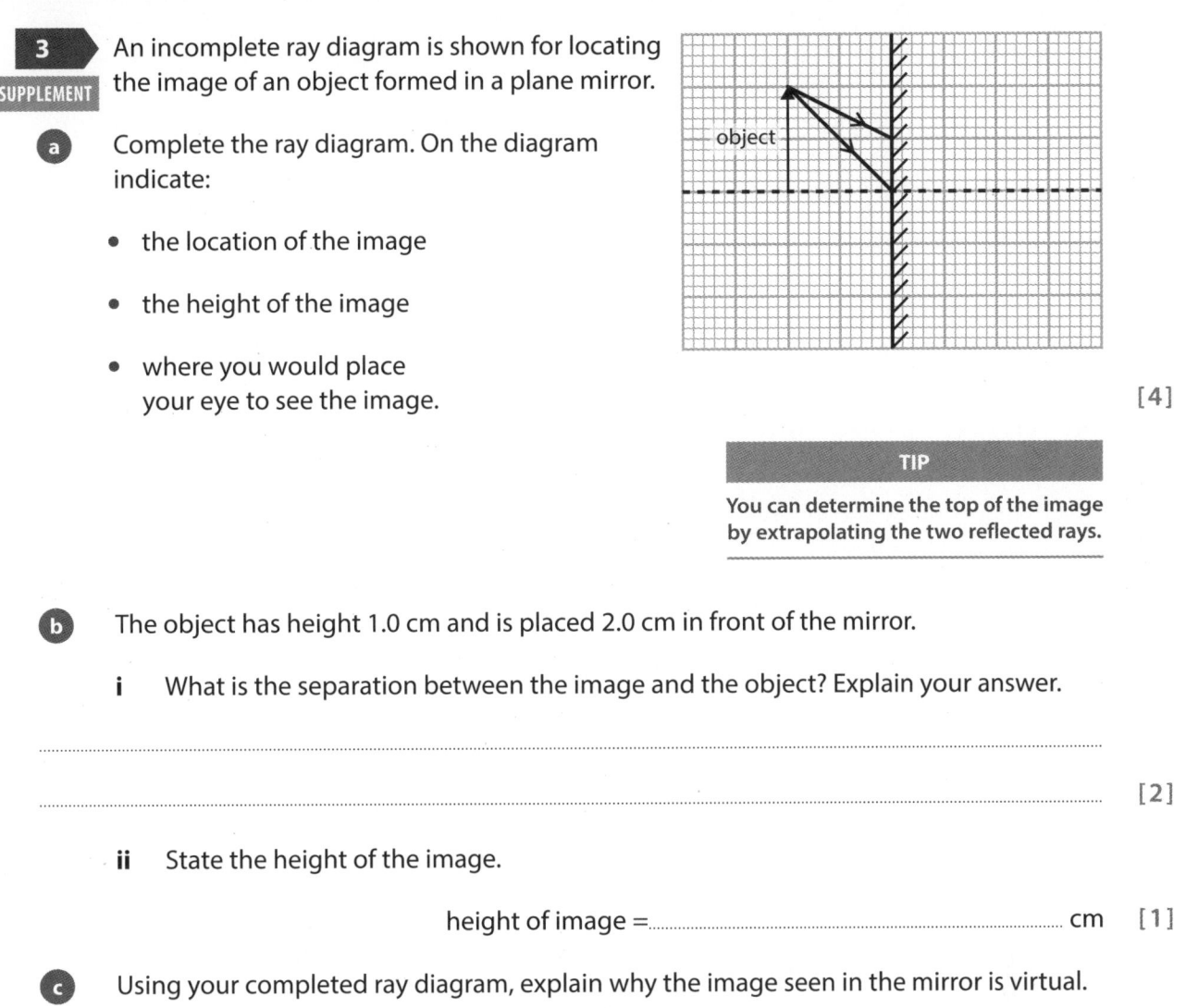

TIP

You can determine the top of the image by extrapolating the two reflected rays.

b The object has height 1.0 cm and is placed 2.0 cm in front of the mirror.

  i What is the separation between the image and the object? Explain your answer.

  ....

  .... [2]

  ii State the height of the image.

  height of image = ... cm [1]

c Using your completed ray diagram, explain why the image seen in the mirror is virtual.

....

.... [2]

181

Refraction of light

Student's Book pages 573–574 | Syllabus learning objectives P3.2.2.1–P3.2.2.3

1 Complete the sentence below.

Refraction is the change in .... of light as it passes from

one .... to another. [2]

2 The diagram below shows a ray of light from a laser that is incident at the surface of a rectangular glass block. The ray emerging from the block is also shown.

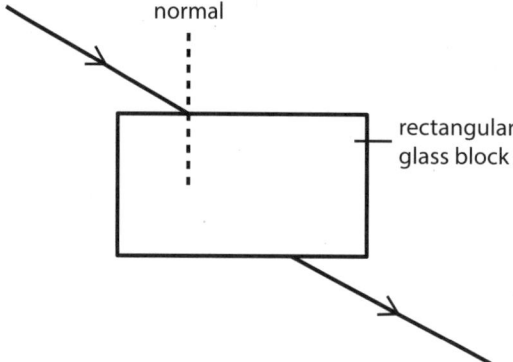

The diagram is drawn to scale.

a Draw the refracted ray within the glass block. [1]

b Use the diagram to measure the angle of incidence i and the angle of refraction r using a protractor.

i = ... degrees r = ... degrees [2]

3 A ray of light enters a triangular glass block placed in air.

Complete the path of the light through the block and as it emerges back into the air.

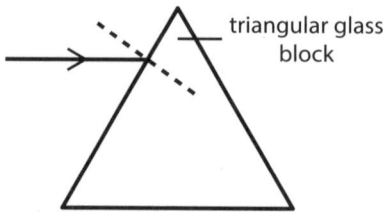

[2]

Lenses and dispersion

Student's Book pages 574–577 | Syllabus learning objectives P3.2.3.1–P3.2.3.4; P3.2.4.1–P3.2.4.2

1 Complete the three definitions below.

a The principal axis of a lens is a line of symmetry passing through the ……………………………… of the lens. [1]

b The principal focus (focal point) of a converging lens is a ……………………………… at which rays of light ……………………………… to the principal axis converge to after passing through the lens. [2]

2 a The diagram opposite shows two rays of light passing through a thin converging lens.

On the diagram show the principal focus (focal point) with a letter F and indicate the focal length f of the lens.

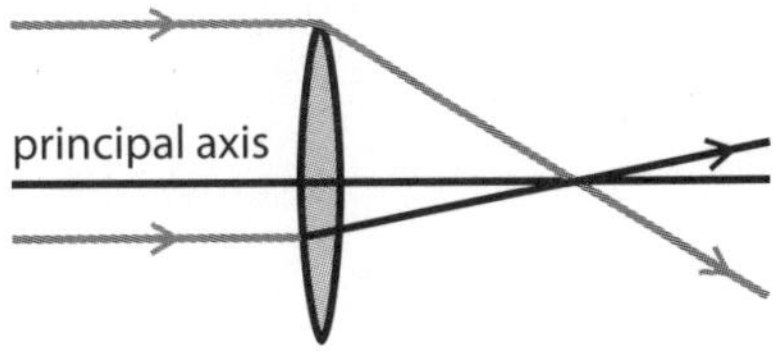

[2]

b The Sun is a very distant object. Almost parallel rays of light from the Sun arrive at the Earth.

You are given a piece of white card and a converging lens of unknown focal length f.

Describe how you could use the white card on a sunny day to determine f.

[3]

3 A thin converging lens is used to form a sharp image of a light bulb (object) on a screen. In the diagram below, the object is shown by an upright arrow and labelled O. The principal focus on either side of the lens is marked by the letter F.

The ray diagram shows the path of two rays of light from the top of the object.

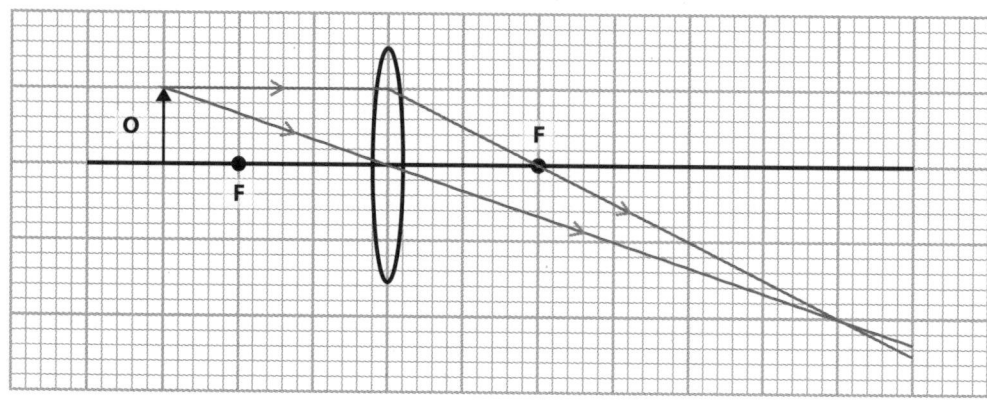

a Draw the position of the image. [1]

b Circle the words below that correctly describe the characteristics of the image formed on the screen.

        diminished   inverted   magnified (enlarged)

                      real   upright [3]

4 a Describe how you can demonstrate the dispersion of white light in the laboratory.

....

.... [2]

b State how the wavelength of yellow light compares with the wavelength of red light.

.... [1]

c List blue, red and green colours of light in order of increasing frequency.

.... [1]

d The dispersion of white light produces seven coloured lights. List the colours of the light in the order of increasing wavelength.

.... [1]

Electromagnetic spectrum

Student's Book pages 581–585 | Syllabus learning objectives P3.3.1–P3.3.4;
SUPPLEMENT P3.3.5

1 This question is about the main regions of the electromagnetic spectrum.

a A student has listed below the main regions of the electromagnetic spectrum in order of **increasing** frequency.

→

radio waves; microwaves; infrared; ultraviolet; visible light; X-rays; gamma rays

Two main regions in the list are not in the correct sequence. What is the error? Suggest how this error can be resolved.

....

.... [2]

b State which main region of electromagnetic waves has the shortest wavelength.

.... [1]

c List all the main regions of electromagnetic waves that have a wavelength longer than that of infrared waves.

.... [1]

d List all the main regions of electromagnetic waves that have a frequency greater than that of visible light.

.... [1]

2 The electromagnetic radiation emitted by a distant exploded star is being observed from the Earth. The star has been known to emit all types of electromagnetic radiation.

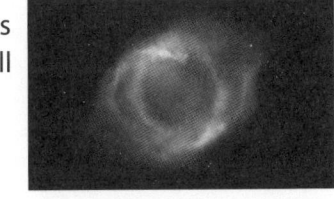

Which statement is correct? Circle the correct letter.

    A All the electromagnetic waves from the star have the same speed in vacuum.

    B The higher frequency electromagnetic waves from the star travel faster in vacuum.

    C Visible light from the star travels through a vacuum faster than the X-rays.

    D The speed of electromagnetic radiation in a vacuum and in air is very different. [1]

P3 Waves | Electromagnetic spectrum

3 SUPPLEMENT

a Which statement is correct about the speed of electromagnetic waves? (Circle) the correct letter.

A Electromagnetic waves cannot travel through a vacuum.

B The speed in a vacuum is 340 m/s.

C The speed in a vacuum is 3.0×10^8 m/s.

D The speed depends on its wavelength. [1]

b Calculate the distance travelled by radio waves in 1 minute.

> **TIP**
> Radio waves are electromagnetic waves.

distance = ... m [3]

c The circumference of the Earth is about 40 000 km.

Calculate the time it would take for light to travel this distance.

time = ... s [3]

4 Electromagnetic waves have many uses.

Each row in the table below shows some of the uses for a particular type of electromagnetic wave.

A	Radio transmission and radar
B	Remote controllers for digital devices and thermal imaging
C	Detecting fake bank notes and sterilising water
D	Medical scanning and security scanners
E	Detecting and treatment of cancer
F	Satellite television and mobile phones (cell phones)
G	Photography and illumination

In the table below, identify the correct row for each type of electromagnetic wave. One has already been done for you.

Electromagnetic wave	The correct row for uses
Radio waves	
Microwaves	
Infrared	
Visible light	
Ultraviolet	
X-rays	
Gamma rays	E

[6]

5 In each case below, identify the type electromagnetic radiation that may have caused the harmful effect in patients being treated at a hospital after excessive exposure to a particular type of radiation.

a Damage to cells in the hand .... [1]

b Mutation of cells in the lung .... [1]

c Skin cancer on the arm .... [1]

Sound and ultrasound

Student's Book pages 588–593 | Syllabus learning objectives P3.4.1–P3.4.7; **SUPPLEMENT** P3.4.8–P3.4.10

1

a Describe a simple method of creating a sound that can be heard.

[1]

b State a typical value for the frequency of the emitted sound.

frequency = ... Hz [1]

2

a Explain why sound created on a distant planet cannot be heard on the Earth, no matter how sensitive our detecting devices are.

[2]

b A person at a distance of 200 m creates a loud sound by banging together two blocks of wood. The speed of sound in air is 340 m/s. The student wishes to directly measure the time it would take for sound to travel 200 m using a stopwatch.

   i Calculate the time it would take sound to travel this distance.

time = ... s [3]

   ii Suggest why it would not be sensible to measure the time taken for sound to travel 200 m with a stopwatch.

[1]

P3 Waves | Sound and ultrasound

3 A scientist is investigating the speed of sound in air and in seawater. A fixed distance of 1.0 km is used. The scientist records the time t taken for the sound waves to travel this distance.

SUPPLEMENT

The experiment is repeated four times for both air and seawater. The table below shows the data collected.

t (s) for air	3.00	2.95	3.08	2.90	
t (s) for seawater	0.67	0.72	0.66	0.70	

a Determine the average times for air and seawater. Write the values, to two decimal places, in the last column. [2]

b Use the information above to deduce whether the speed of sound is air or in seawater.

....

.... [2]

4 A loudspeaker is placed next to a wall. There is another wall opposite at a distance of 6.0 m. The loudspeaker creates a single loud bleep. The sound waves are reflected at the walls.

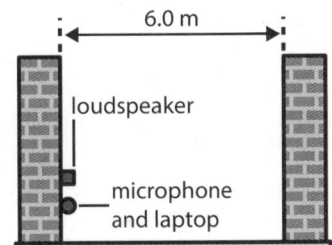

A microphone is placed next to the loudspeaker. A laptop connected to the microphone shows the following pattern for the sound detected.

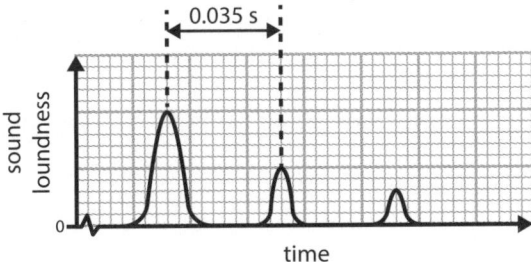

a What is another term for the reflection of sound?

.... [1]

b Explain why several 'peaks' are detected by the microphone.

....

.... [2]

c The time between the adjacent peaks is 0.035 s.

　i Explain why the distance travelled by the sound is 12 m in this time.

.... [1]

　ii Calculate the speed of sound.

> **TIP**
> Use the information provided in (c)(i) to calculate the speed.

speed = ... m/s [2]

d State how you could make the initial bleep from the loudspeaker:

　i louder

.... [1]

　ii have a higher pitch.

.... [1]

5 Ultrasound is high-frequency sound waves that humans cannot hear.

State a typical frequency for ultrasound in kHz. .... [1]

6 The speed of sound in a particular liquid is about 1200 m/s.

SUPPLEMENT

a State if the speed of sound in a gas (e.g. air) is less than, or the same as, or more than this value.

.... [1]

b State if the speed of sound in a solid (e.g. iron) is less than, or the same as, or more than this value.

.... [1]

c Sound waves are longitudinal and consist of a series of compressions and rarefactions.

Describe the nature of a compression and rarefaction regions of a sound wave in terms of pressure and how close the particles are.

....

.... [2]

P4 Electricity
Electrical charge

Student's Book pages 600–604 | Syllabus learning objectives P4.1.1.1–P4.1.1.3; **SUPPLEMENT** P4.1.1.4

1 Electric charge can either be positive (+) or negative (−). The forces between charged objects can either be attractive or repulsive.

a In the diagram below, state whether there is an attractive force, or a repulsive force, between the charged objects.

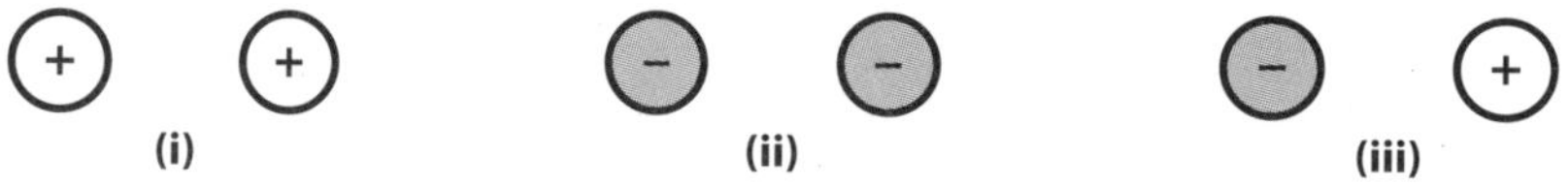

(i) .... (ii) .... (iii) .... [3]

b Formulate a rule for the type of force using the terms 'like charges' and 'unlike charges'.

.... [1]

2 The simple circuit opposite is used by a student to identify whether a material is an electrical conductor, or an electrical insulator. The 'test' material is placed between the contacts P and N.

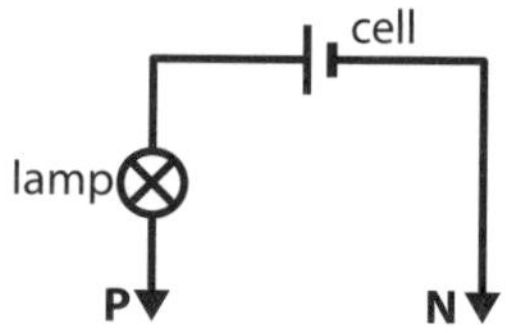

a Explain what happens to the lamp:

i when an insulator is placed between the contacts

....

.... [2]

ii when a conductor is placed between the contacts.

....

.... [2]

b In the list below, circle the **three best** electrical conductors.

aluminium   copper   gold   oil   plastic   silk   wood   [1]

3 **SUPPLEMENT** What is the unit for electric charge? Circle the correct letter.

   A   ampere   B   coulomb   C   hooke   D   newton   [1]

Electric current

Student's Book pages 604–607 | Syllabus learning objectives P4.1.2.1–P4.1.2.4; **SUPPLEMENT** P4.1.2.5–P4.1.2.7

1 a Some students confuse electric charge and electric current. These quantities are related to each other, but they are not the same.

Electric current is related to the flow of which quantity? Circle the correct letter.

   A   charge   B   energy   C   force   D   time   [1]

b Which particles move in the copper wire of an electric circuit when there is conduction (or a current)? Circle the correct letter.

   A   atoms   B   electrons   C   ions   D   protons   [1]

c State one instance in which a digital ammeter would be preferable to an analogue ammeter.

.... [1]

2 The three graphs below show the variation of electric current with time.

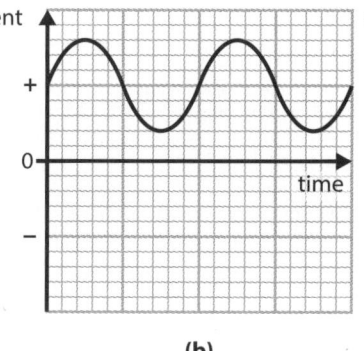

(a)   (b)   (c)

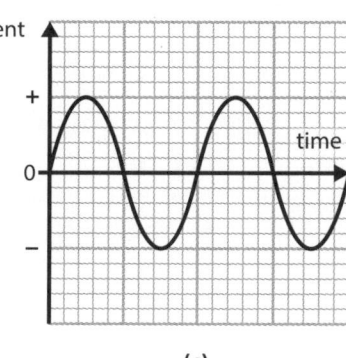

In each case, state whether the current is direct or alternating. Give a reason for each answer.

(a) ....

....

(b) ....

....

(c) ....

.... [6]

3 SUPPLEMENT

a Complete the definition for electric current below.

Electric current is defined as the .... that passes a point in a circuit .... unit time. [1]

b In the circuit opposite, the ammeter shows a constant reading of 0.50 A.

i State the **relationship** between the charge Q and the time t taken for this charge to flow past a point in a circuit.

....

.... [2]

ii Calculate the charge Q moving past any point in the circuit in a time of 20 s.

$Q = $ .... C [3]

iii Using either the term 'clockwise' and/or the term 'anticlockwise', state the direction of:

- the flow of delocalised (mobile) electrons in the circuit

.... [1]

- the conventional current.

.... [1]

c Calculate the current I in milliamperes (mA) in a battery charger when the charge flow in a time of 60 s is 15 C.

> **TIP**
> The prefix milli means 1000 times smaller. So, 1 A = 1000 mA.

$I = $ .... A [3]

Electromotive force and potential difference

Student's Book pages 608–609 | Syllabus learning objectives P4.1.3.1–P4.1.3.3;
SUPPLEMENT P4.1.3.4–P4.1.3.7

1 a Name a meter that can be used to measure the voltage across a component in an electrical circuit.

.... [1]

b A student is measuring the voltage across a length of wire. The circuit diagram drawn by the student is shown opposite.

What is the mistake made by the student? Suggest how you can correct this mistake.

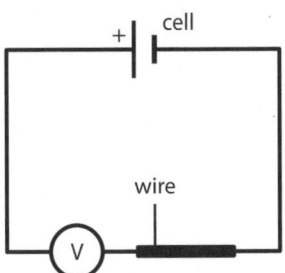

....
.... [2]

c Two components are connected in series to a 6.0 V battery.

　i Name the device that provides the current in this circuit.

.... [1]

ii The voltage across one of the components is 2.0 V. State the voltage across the other component. Explain your answer.

....

.... [2]

d Which statement is correct about electromotive force (e.m.f.)? Circle the correct letter.

SUPPLEMENT

A It is related to the force experienced by charges in a complete circuit.

B It is the electrical work done in moving a unit charge in a complete circuit.

C It is the total current in a complete circuit.

D It is the work done by the complete circuit. [1]

e Circle all the items in the list below that are commonly used to define **both** electromotive force (e.m.f.) and potential difference (p.d.).

SUPPLEMENT

current   force   unit charge   unit mass   unit volume   work done

[2]

2 A small heater consists of a coiled copper wire. This heater and a lamp are connected in series to a power supply. The electromotive force (e.m.f.) of the power supply is 12 V. The current in the heater is 2.0 A.

The potential difference across the lamp is 4.0 V.

a Calculate the potential difference (p.d.) across the heater.

potential difference = ... V [2]

b Show that the charge flow in the heater in 120 s is 240 C.

SUPPLEMENT

> **TIP**
> You must clearly show all the steps of the calculations because the answer is given.

[2]

c **SUPPLEMENT** The e.m.f. of the power supply is 12 V. State the electrical work done by the power supply in moving a unit charge (1 C) around the complete circuit.

.... J   [1]

d **SUPPLEMENT** State the work done by a unit charge (1 C) passing through the heater.

.... J   [1]

Resistance

Student's Book pages 609–614 | Syllabus learning objectives P4.1.4.1; **SUPPLEMENT** P4.1.4.2

1 A student is given the task of determining the resistance of a pencil lead. The ends of the pencil are exposed as shown.

The ends of pencil lead are connected to a cell. The current I is measured using an ammeter and voltage V across the pencil lead is measured using a voltmeter.

State the formula you would use to calculate the resistance of the pencil lead.

.... [1]

2 The table opposite shows some data collected by a student investigating the resistance of an electrical component.

V (V)	2.0	4.0	6.0
I (A)	1.8	2.5	3.2
R (Ω)			

a Complete the table by determining the resistance of the component. Write each calculated value to two significant figures. [3]

b Use the completed table to suggest the variation of R with V.

.... [1]

3 a The circuit shown opposite is used by a teacher to demonstrate how the resistance of a wire of a given material depends on its length.

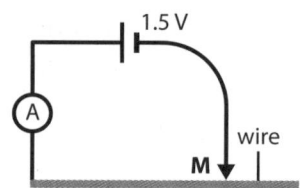

As the movable contact M is moved to the right, the current in the circuit slowly decreases.

Explain how the resistance of the wire depends on its length.

....

.... [2]

b The table below has some statements about a wire made from nickel.

Insert a tick ✓ in the last column if the statement is correct. [4]

Statement	Place your tick ✓ here if correct
A thicker wire has less resistance.	
A shorter length of wire has less resistance.	
The resistance of wire decreases as the current in it is increased.	
The resistance of wire increases as the potential difference is decreased.	

4 a The resistance of 0.90 m long nichrome-wire of a given cross-sectional area has resistance of 18 Ω.

Calculate the resistance R of a 0.60 m length of this wire.

$R =$ .... Ω [3]

b Calculate the resistance R of 0.90 m long nichrome-wire that has half the cross-sectional area of the wire in **(a)**.

$R =$ .... Ω [3]

Energy and power

Student's Book pages 614–617 | Syllabus learning objectives P4.1.5.1–P4.1.5.4

1 A simple circuit consists of a cell that is connected in series with a resistor and a filament lamp.

a State the source of the electrical energy.

.... [1]

b Name one component that energy is transferred to by the circuit.

.... [1]

c The components connected to the cell may become warm. In such situations, where is the internal (thermal) energy of the components transferred to? Circle the correct letter.

    A the cell   **C** the meters

    B the connecting wires   **D** the surroundings   [1]

2 A lamp is rated as 2.3 W.

> **TIP**
> Remember 1 W = 1 J per second.

a State the energy transferred by the lamp in a time of 1.0 s.

.... [1]

b Calculate the energy transferred by the lamp when it is operated for 1.0 hour.

> **TIP**
> There are 60 minutes in 1 hour, and 60 seconds in 1 minute.

energy transferred = ... J [2]

c The lamp is connected to a 230 V supply. Calculate the current in the lamp.

current = ... A [3]

P4 Electricity | Energy and power

3 An appliance of resistance 4.0 Ω is connected to a 12 V supply.

a Calculate the current I in the appliance.

$I =$ .... A [3]

b Calculate the power P transferred by the appliance.

$P =$ .... W [3]

c The kilowatt-hour (kW h) is an alternative unit for energy.

  i Define the kW h.

  ....

  .... [1]

  ii The cost of each kW h is 18 cents. Calculate the cost of operating the appliance for 12 hours.

  cost = ... cents [3]

d Explain whether a 60 W device operated for the same period as the appliance, with the same supply, would be more costly, the same cost, or less costly.

....

.... [2]

Circuit diagrams and components

Student's Book pages 622–623; 629–630 | Syllabus learning objectives P4.2.1.1; **SUPPLEMENT** P4.2.1.2

1 a) A simple circuit consists of a fixed resistor, an ammeter and a switch all connected in series to a battery.

Using the correct symbols for the components, draw a diagram for this circuit.

[5]

b) The circuit in **(a)** is modified by adding a lamp in the circuit and replacing the fixed resistor with a variable resistor.

Draw the circuit symbols for a variable resistor and a lamp.

Circuit symbol for a variable resistor	
Circuit symbol for a lamp	

[2]

c) Draw a circuit symbol for a generator.

[1]

2 This question is about identifying electrical components in a circuit. You are not expected to have knowledge of this circuit.

Identify the components A, B, C, D and E.

A: ....

B: ....

C: ....

D: ....

E: ....

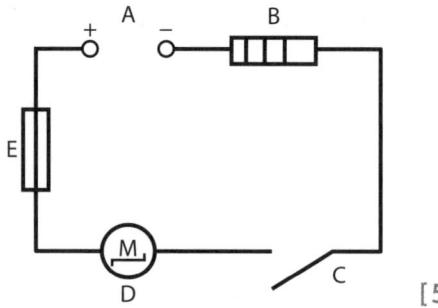

[5]

3 A diode is a component that conducts in one direction only. A light emitting diode (LED) is a special diode that emits light of a particular colour when it is conducting.

SUPPLEMENT

The circuit below is designed by a student as a simple polarity-checker for cells, batteries and power supplies.

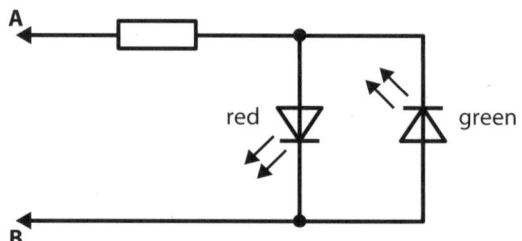

a State which LED is lit when A is positive and B is negative.

.... [1]

b State which LED is lit when A is negative and B is positive.

.... [1]

Series and parallel circuits

Student's Book pages 623–628 | Syllabus learning objectives P4.2.2.1–P4.2.2.6;
SUPPLEMENT P4.2.2.7–P4.2.2.8

1 A row of lamps, used to illuminate a courtyard, are connected in parallel rather than in series.

Which statement is correct about the lamps connected in parallel? Circle the correct letter.

    A Lamps connected in parallel last longer.

    B Lamps give out less light when connected in parallel.

    C Parallel circuits are easier to connect.

    D When one lamp stops working the remaining lamps are unaffected. [1]

2 The circuit opposite has three identical filament lamps connected in series.

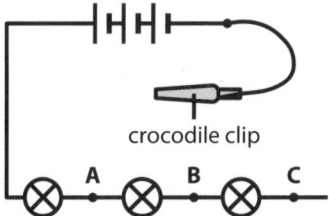

One end of the connecting wires has a crocodile clip. This clip can be connected to any of the points A, B or C.

The electromotive force (e.m.f.) provided by the battery of three cells is 4.5 V.

a The crocodile clip is connected to point C. Explain why all the lamps have the same brightness.

....

.... [2]

b The crocodile clip is now moved to point B.

The current in the circuit is 0.030 A. The brightness of the two lamps is **brighter** than the lamps in **(a)**.

    i State whether the current in the circuit is more than, the same as, or less than the current in the circuit in **(a)**.

.... [1]

    ii Explain whether the circuit in **(a)** or the circuit in **(b)** has greater resistance.

....

.... [2]

iii Calculate the total resistance of the two lamps.

total resistance = ... Ω [3]

iv Calculate the resistance of each lamp. Show all the steps in your calculation or reasoning.

resistance of each lamp = ... Ω [3]

3 This question is about the circuit opposite. The resistors have different resistance values. One of resistors has resistance R. The electromotive force (e.m.f.) of the power supply is 9.0 V.

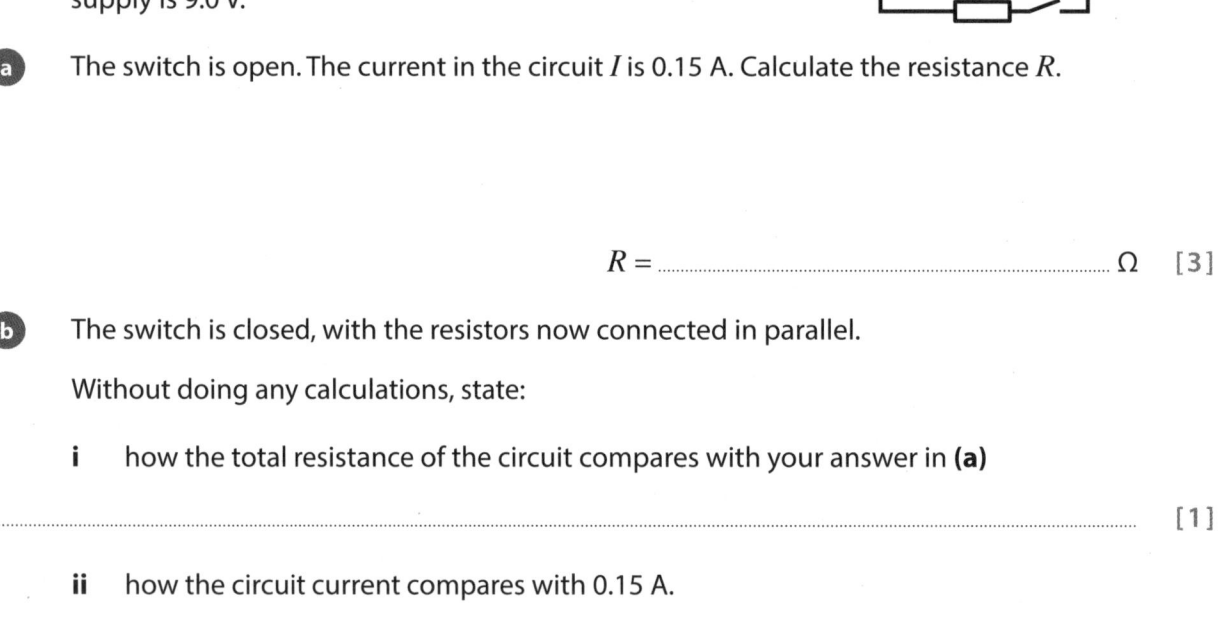

a The switch is open. The current in the circuit I is 0.15 A. Calculate the resistance R.

$R = $ .... Ω [3]

b The switch is closed, with the resistors now connected in parallel.

Without doing any calculations, state:

i how the total resistance of the circuit compares with your answer in **(a)**

.... [1]

ii how the circuit current compares with 0.15 A.

.... [1]

4 Two lamps are connected in parallel across the terminals of a battery.

The current supplied by the battery is 5.0 A.

In the list below, circle the current that is **impossible** in either of the lamps

1.2 A   5.4 A   3.1 A   4.7 A   2.5 A   [1]

5 a For each circuit, calculate the total resistance R_C of the circuit.

i 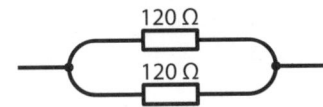   ii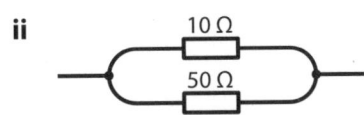

R_C for (i) = ... Ω R_C for (ii) = ... Ω   [5]

b Complete the following sentence that is applicable to both circuits in (a).

The total resistance is always .... than the smaller resistance of the two resistors. [1]

6 a This question is about the circuit shown opposite.

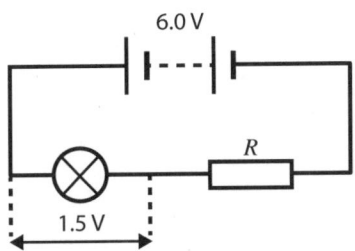

The e.m.f. of the battery is 6.0 V. The potential difference (p.d.) across the lamp is 1.5 V. The current in the circuit is 0.015 A.

i Show that the p.d. across the fixed resistor is 4.5 V.

[1]

ii Calculate the resistance R of the fixed resistor.

> **TIP**
> The current in a series circuit is the same.

R = ... Ω   [3]

b This question is about the circuit opposite. The values of the resistors and the current in one of the resistors are shown on the circuit diagram.

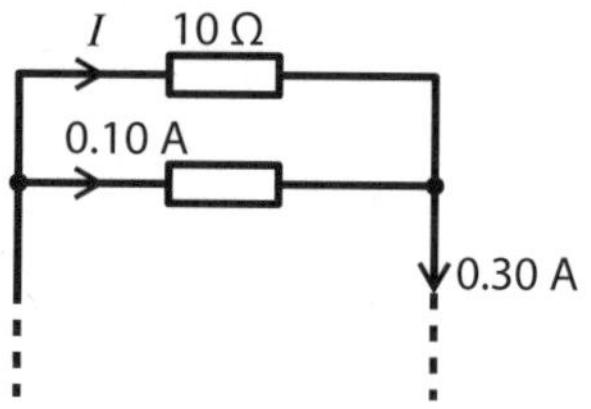

  i Determine the current I in the 10 Ω resistor.

  $I = $ .... A [1]

  ii Calculate the p.d. across the parallel combination. Explain your answer.

  p.d. = ... V [3]

Electrical safety

Student's Book pages 635–639 | Syllabus learning objectives P4.3.1–P4.3.4

1 In many countries the mains voltage is between 120 V and 230 V. Any contact with such high voltages can be extremely dangerous.

a State why it may be sensible to have light switches outside a bathroom.

.... [1]

b Explain the danger of having an appliance cable as shown opposite.

.... [2]

c Explain why it may not be sensible to use a very thin cable for an electric kettle.

.... [2]

d What is the likely risk of using too many appliances connected to an extension cable? Give a reason for your answer.

....

.... [2]

2 A mains plug will always have a fuse inside. It is important that the fuse has the correct rating.

a Explain the purpose of a fuse.

....

.... [2]

b The diagram shows part of the label for an electric kettle.

LISTED 8H45
TYPE K15 KETTLE
120 V 60 Hz 1500 W

i Determine the operating current for this kettle.

current = ... A [3]

ii Which of the following fuses would be suitable in the plug of the electric kettle? Circle the correct letter. Give a reason for your answer.

5.0 A   13 A   20 A

....

.... [2]

3 A fuse wire is an important safety device in a mains plug.

Here are five **incorrectly** sequenced statements about the operation of a fuse.

1. A fault occurs in the appliance.
2. The appliance switches off.
3. The fuse wire melts.
4. The fuse breaks the live connection to the appliance.
5. There is a large current in the fuse wire.

Place the statement numbers in the correct sequence.

.... [1]

P4 Electricity | Cells, batteries, generators and motors

4 **a** A double-insulated appliance is connected by a cable to the mains supply. It has two layers of insulation (plastic) surrounding the live wire. State why the earth wire does not need to be connected to the plug for this appliance.

....

.... [1]

b A row of trip switches in a building are shown opposite.

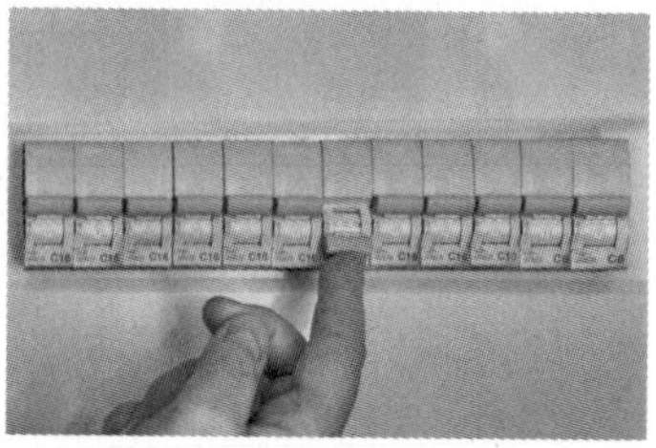

Which statement below applies to a trip switch? Circle the correct letter.

A It is a circuit that sounds an alarm when the current is too large.

B It is a device that switches off when there is excessive current.

C It is a short length of wire that melts when the current is too large.

D It is a simple on-off mechanical switch. [1]

c A different appliance with a metal casing has the earth connection to the metal case snapped off. Describe why this would be dangerous to the user when the live wire accidently touches the casing.

....

.... [2]

Cells, batteries, generators and motors

Student's Book pages 642–643 | Syllabus learning objectives P4.4.1-P4.4.3

1 **a** Name an electrical device that transfers electrical energy into kinetic energy.

.... [1]

b Describe the energy transfers for a generator.

.... [1]

c Complete the sentence below.

Batteries and cells transfer chemical energy into .... energy. [1]

208

P5 Space physics
The Solar System

Student's Book pages 650–652 | Syllabus learning objectives P5.1.1.1

1 This question is about the Solar System.

a Name the four planets that orbit the closest to the Sun.

[1]

b Name the four planets that orbit the farthest from the Sun.

[1]

c Describe how the structure (size and composition) of the planets mentioned in **(a)** differ from the planets mentioned in **(b)**.

[2]

2 Here is a list of terms that **may** be the answers to the questions that follow.

| Saturn | Sun | Jupiter | Mars |
| Mercury | moons | Neptune | Pluto |

a Which two planets come closest to Uranus in their orbits?

[1]

b Name a dwarf planet.

[1]

c Name the two planets between which **most** of the objects in the asteroid belt are located.

[1]

d Name the star of our Solar System.

[1]

e Small objects orbits around Jupiter. What do astronomers call these objects?

[1]

Stars

Student's Book pages 654–662 | Syllabus learning objectives P5.2.1.1–P5.2.1.5, P5.2.2.1–P5.2.2.2; **SUPPLEMENT** P5.2.1.6–P5.2.1.8; P5.2.2.3

1 The Sun is the closest star to Earth.

a The Sun radiates its energy in the form of electromagnetic radiation. In the list below, circle the three regions of the electromagnetic radiation where it emits **most** of its energy.

   radio waves   infrared   visible   ultraviolet   X-rays   [1]

b The Sun is stable star that radiates energy into space.

SUPPLEMENT Describe how energy is produced in the Sun.

.... [2]

2 The nearest star to the Sun is Proxima Centauri. It is about 4.25 light-years away.

a Define the term light-year (ly).

.... [1]

b The speed of light in a vacuum is 3.0×10^8 m/s and 1 year $= 3.15 \times 10^7$ s. Calculate the distance of 4.25 ly in metres (m).

4.25 ly = ... m [3]

c Suggest why light-year is used to measure the distance of stars and galaxies.

.... [1]

d The centre of our galaxy is about 30 000 ly away. How long does it take for light to travel from the centre of the galaxy to us?

.... [1]

3 This question is about planets orbiting the Sun.

a Here are some facts about Jupiter written by a student.

- It is the most massive planet in the Solar System.
- It has mass that is about 2.5 times the total mass of all the planets in the Solar System.
- The mass of the Sun is more than 1000 times the mass of Jupiter.

Suggest why the planets in the Solar System orbit the Sun.

.... [1]

b Describe the nature of the force that keeps a planet in its orbit around the Sun.

....

.... [2]

4 The speed of light in a vacuum is 3.0×10^8 m/s.

Calculate the time it takes for light to travel from Mars to the Earth when the distance between them is 78 million km.

> **TIP**
> You need to convert the distance into metres (m).

time = ... s [3]

5 Here are some data on Mars and Neptune.

SUPPLEMENT

Planet	Orbital distance from Sun (10^6 km)	Orbital duration or period (Earth days)
Mars	227.9	687.0
Neptune	4495.1	59 800

The orbital speed v of a planet can be calculated using the equation $v = \dfrac{2\pi r}{T}$

a State what r and T represent.

.... [1]

b Calculate the orbital speed v in m/s of the planets Mars and Neptune.

TIP
1 day = 24 × 3600 = 86 400 s

v (**Mars**) = ... m/s [3]

v (**Neptune**) = ... m/s [3]

c State which of these two planets experience the greater strength of the Sun's gravitational field. Explain your answer.

....

.... [2]

6 This question is about the formation and life cycle (evolution) of stars.

a Describe how a stable star is formed from an interstellar cloud of gas and dust.

....

.... [2]

b Describe how stable stars are powered by nuclear reactions.

....

.... [2]

c A supernova is a catastrophic event in which a star implodes rapidly.

What is created during this event that may be responsible for forming new stars and planets in the future? Circle the correct letter.

A black holes

B heavier elements

C radiation

D white dwarfs [1]

d Here is a list of some objects.

   black hole   neutron star   planetary nebula   red giant

   red supergiant   star   supernova   white dwarf

Use the objects from the list to:

   i describe the life cycle of a small mass star (e.g Sun)

   ....

   ....

   .... [3]

   ii describe the life cycle of a large mass star.

   ....

   ....

   .... [3]

Galaxies and the Universe

Student's Book pages 662–663 | Syllabus learning objectives P5.2.3.1–P5.2.3.2; **SUPPLEMENT** P5.2.3.3

1 a What is the approximate diameter of our galaxy (Milky Way)? Circle the correct letter.

   A 1000 ly   **B** 10 000 ly   **C** 100 000 ly   **D** 1 000 000 ly   [1]

b List the following objects in the order of **increasing** distance from the Earth.

   Andromeda galaxy   the Moon   another star in the Milky Way   the Sun

   .... [1]

2 It is estimated that there are 100 billion galaxies in the Universe. Each galaxy has about 100 billion stars.

TIP
1 billion is 1 000 000 000 or 10^9.

213

a Determine how many stars there are in the Universe.

number of stars = ... [2]

b The average mass of a star is about 2.0×10^{30} kg.

Estimate the total mass of matter in the Universe.

mass = ... kg [2]

3 **a** Describe what is meant by the Big Bang Theory.

SUPPLEMENT

....

....

.... [3]

b What is the state of the Universe? Circle the correct letter.

A It is stationary.

B It is contracting.

C It is expanding.

E It is getting younger. [1]

c What is the approximate age of the Universe? Circle the correct letter.

A 13.8 seconds

B 13.8 years

C 13.8 million years

D 13.8 billion years [1]